Low Earth Orbital Satellites for Personal Communication Networks

For a complete listing of the *Artech House Mobile Communications Library,*
turn to the back of this book.

Low Earth Orbital Satellites for Personal Communication Networks

Abbas Jamalipour

Artech House
Boston • London

Library of Congress Cataloging-in-Publication Data

Jamalipour, Abbas

    Low earth orbital satellites for personal communication networks / Abbas Jamalipour
        p.   cm.—(Artech House mobile communications library)
    Includes bibliographical references and index.
    ISBN 0-89006-955-7 (alk. paper)
    1. Artificial satellites in telecommunication.   2. Mobile communication systems.
      I. Title.   II. Series: Artech House telecommunications library.
TK5104.J35   1997
621.3845—dc21

                                                        97-32244
                                                        CIP

British Library Cataloguing in Publication Data

Jamalipour, Abbas

    Low earth orbital satellites for personal communication networks
    1. Artificial satellites in telecommunication
    I. Title
    621.3'8254

    ISBN 0-89006-955-7

Cover and text design by Darrell Judd.

© 1998 ARTECH HOUSE, INC.
685 Canton Street
Norwood, MA 02062

International Standard Book Number: 0-89006-955-7
Library of Congress Catalog Card Number: 97-32244

10 9 8 7 6 5 4 3 2

Contents

Preface   ix

    Acknowledgments   xi

Introduction   xiii

    Organization of this book   xv

1   Mobile Satellite Communications   1

    1.1   Communications satellites   2

      1.1.1   Preliminary issues   2

      1.1.2   History of communications satellites   5

    1.2   Orbital dynamics of satellite systems   7

      1.2.1   Kepler's first law   8

      1.2.2   Kepler's second law   8

      1.2.3   Kepler's third law   8

      1.2.4   An example: The geostationary orbit   10

1.3   Mobile satellite communications systems   12

   1.3.1   Orbit selection   12

   1.3.2   Mobile satellite systems   18

1.4   Summary   27

2   Communications with LEO Satellites   33

2.1   Preliminary issues in LEO satellite systems   35

   2.1.1   Required number of LEO satellites and orbits   35

   2.1.2   Hand-off   41

   2.1.3   Intersatellite links   43

   2.1.4   Spot beams   46

   2.1.5   Doppler shift effect   51

2.2   Specific issues in LEO satellite systems   55

   2.2.1   Selection of a multiple-access scheme   56

   2.2.2   Traffic considerations   64

2.3   Modeling the LEO satellite systems   67

2.4   Summary   72

3   Application of CDMA in LEO Satellite Systems   77

3.1   Performance evaluation of analog systems   79

   3.1.1   Traffic modeling   79

   3.1.2   SIR: The measure of performance   82

   3.1.3   Traffic assignment control   91

3.2   Performance of integrated voice/data systems   96

   3.2.1   System considerations   96

   3.2.2   Extension of the traffic model   99

   3.2.3   Simulation environment   101

   3.2.4   Performance measurement   103

   3.2.5   Dynamic nonuniform traffic concepts   108

3.3   Summary   113

4   Spread-Slotted Aloha for LEO Satellite Systems   117

4.1   Spread-slotted Aloha   119

4.1.1   The Aloha multiple-access scheme   119

4.1.2   Spreading the Aloha packets   123

4.2   Employing spread-slotted Aloha in a LEO
        satellite system   130

4.2.1   Distribution of users   132

4.2.2   Throughput analysis   134

4.2.3   Probability of packet success   136

4.3   Numerical examples   144

4.4   Summary   151

5   Modified Power Control in Spread-Slotted Aloha   157

5.1   Worst case in throughput performance   159

5.1.1   Intracell interference versus intercell interference   160

5.1.2   Performance of nonworst cases   164

5.2   Modified power control scheme   168

5.2.1   Purpose and structure of the scheme   169

5.2.2   Numerical examples   172

5.2.3   Some practical notes on realization of the scheme   179

5.3   Summary   180

**6   Transmit Permission Control Scheme for
       Spread-Slotted Aloha   185**

6.1   Transmit permission control scheme:
        Nonfading channel   187

6.1.1   Basic considerations   188

6.1.2   Transmit permission control   190

6.1.3   Throughput performance of transmit permission control   195

6.1.4   Average delay performance of transmit permission
          control   202

6.2   Transmit permission control scheme: Fading channel   209

6.2.1   Fading channel model and analysis   209

6.2.2   Numerical examples of the performance of the system   212

6.3   Adaptive transmit permission control schemes   214

   6.3.1   ATPC method 1   215

   6.3.2   ATPC method 2   217

   6.3.3   Performance of ATPC methods   217

6.4   Summary   219

7   Further Considerations in LEO Satellite Systems   223

7.1   Packet admission control scheme   224

   7.1.1   System and traffic models   225

   7.1.2   Evaluation of heavy-traffic performance   228

   7.1.3   Concepts of the scheme   232

   7.1.4   Performance of the scheme   233

7.2   Power control   236

   7.2.1   The near-far problem   236

   7.2.2   Implementation of power control   237

   7.2.3   Effects of imperfections in power control   238

7.3   Multibeam LEO satellites   241

   7.3.1   General expression for antenna gain   242

   7.3.2   Spot-beam antenna gain   243

   7.3.3   Performance of spot-beam antennas   245

7.4   Concept of adaptive array antennas   248

7.5   Summary   251

List of Acronyms   257

About the Author   261

Index   263

Preface

I N THE PAST FEW YEARS, there has been a rush of research toward the realization of a global personal communication network that can provide reliable, ubiquitous, and cost-effective communication services to individuals via small and single-standard hand-held terminals. That trend is expected to continue through the first decade of the next century. The exponential increase in the number of subscribers for mobile telephones during the last five years, as well as increasing trends for multimedia communications, is driving the future of mobile communication systems.

To meet the communication requirements in the upcoming century, global *personal communication networks* (PCNs) have become one of the hottest topics in the field of communications. An important and fundamental question in such plans is which system meets all those requirements. Current cellular systems, although they have good potential for providing voice and data communications in urban areas, would not be a proper choice for a global system. On the other hand, in accordance with the research for satellite communication systems, there has been a widespread desire to set up a single global communication system by satellites, which may be the only solution for the globalization of communications networks.

Thus, *low Earth orbital* (LEO) satellites seem to have some properties over conventional geostationary satellites that make them appropriate candidates for establishing PCNs on a global basis. LEO satellites, while having the important features of conventional geostationary satellites, such as wide coverage area, direct radio path, and flexibility of the network architecture, provide some additional fundamental advantages for global communication networks, for example, short propagation delay, low propagation loss, and high elevation angle in high latitudes.

In recent years, the literature as well as the industry have paid much attention to the commercial use of LEO satellites for establishing a global PCN. Although the history of research on the application of LEO satellites goes back to the early 1960s, the realization of using such satellites on low-altitude orbits for PCN applications is in its infancy. A search through the literature that yielded only a small number of written materials related to this important part of future global communication service prompted me to write this book.

This book is a theoretical study of some of the problems related to the use of LEO satellites for a global communication service. Throughout this book, the reader will find different aspects of the problems that should be considered during the design of any LEO satellite communication networks, as well as a number of references to those systems that cannot be found in the literature so easily. I believe engineers and students can use the contents of this book to start working on LEO satellite systems, but the materials should be modified during the practical realization of LEO satellite systems and according to collected statistics. In that manner, this book will be much more useful if it is used in conjunction with up-to-date technical papers containing practical data of real LEO satellite systems.

This book presents an analytical framework to study the performance of LEO satellite systems, and several problems related to employing those systems in a global PCN are discussed. A major part of the book focuses on the performance of LEO satellite systems when they employ one of two promising multiple access candidates: *code division multiple access*

(CDMA) or spread-slotted Aloha. Another major viewpoint of this book is the problem of nonuniform distribution of the traffic loads around the world, which should be serviced by the LEO satellite system in a global PCN, and is considered here as an original point of view in the LEO satellite systems.

Chapters 1 and 2 are an introduction to the satellite communications system theory as a bridge from the conventional geostationary satellites to the LEO satellites. Some general issues in satellite systems, especially LEO satellite systems, are introduced and these two chapters can be used as an introductory course in satellite systems. The rest of the book presents special analyses for the LEO satellite systems and hence is useful in an advanced or graduated course about LEO satellite systems. The latter part is very much related to spread spectrum techniques. Several excellent textbooks on spread spectrum and CDMA are available, so I do not provide all the fundamentals here. The text, nevertheless, is self-contained: any significant results are derived in the text. Still, to understand Chapters 3 through 7, the reader should have at least an undergraduate electrical engineering background with some probability and communication engineering content. As a text for a graduate-level course, the book can be covered in one semester or, with some compromises, even in one quarter.

Acknowledgments

I would like to express my heartfelt gratitude to my colleagues at Nagoya University, where I did most of the analyses. Prof. A. Ogawa established a good environment and helped me a lot during my stay at the university. Prof. R. Kohno from Yokohama National University, in Japan, Prof. B. Vucetic from Sydney University, in Australia, and many others encouraged me during the writing of this book; I express my sincere thanks to all of them. Many parts of this book have been published in international journals. I acknowledge the constructive comments from the anonymous reviewers of the IEEE, the IEICE, and others that helped me improve those papers and thus this current book. The reviewers at Artech House also provided many useful comments and suggestions, most of which have been incorporated in the book. I hope that the materials given here can help designers of future LEO satellite PCNs design reliable and realistic systems and that we see the first commercial stage of a global communication network provided by the LEO satellites soon.

A. Jamalipour
Nagoya University, Japan
1998

Introduction

THE ESTABLISHMENT of *personal communication networks* (PCNs) on a global basis has recently become one of the hottest topics in the field of communications. Future PCNs are expected to offer reliable, ubiquitous, and cost-effective communication services to individuals via small hand-held terminals, while *low Earth orbital* (LEO) satellite communication systems seem to have properties that make them appropriate for supporting PCNs. Like conventional geostationary satellite systems, LEO satellite systems offer a wide coverage area, a direct radio path, and a flexible network architecture. Unlike their conventional counterparts, however, LEO satellites also provide small propagation delay and loss, and a high evaluation angle at high latitudes.

This book discusses the use of LEO satellite system for a global PCN and the different problems related to that utilization. The discussion focuses on the performance of LEO satellite systems with employment of either CDMA or spread-slotted Aloha. The selection of a multiple access scheme that can efficiently share the limited frequency spectrum to a large number of users is a fundamental issue in any mobile communication system. Another major viewpoint of this book is the problem of nonuniform distribution of the traffic loads around the world, which should be serviced by the LEO satellite system in a global PCN and which is considered here as an original point of view in the LEO satellite systems.

While there does not appear to be a single multiple access technique that is superior to others in all situations, there are characteristics of spread spectrum waveforms that give CDMA certain distinct advantages. The two basic problems that the mobile radio system designer faces are multipath fading of the radio link and interference from other users in the reuse environment. Spread spectrum signals are effective in mitigating multipath because their wide bandwidth introduces frequency diversity. They also are useful in mitigating interference, again because of their wide bandwidth. The result of those effects is a higher capacity potential compared to that of non-spread multiple access methods. Using the spread spectrum techniques in conjunction with the simple conventional slotted Aloha multiple access scheme, namely, spread-slotted Aloha, also results in an interesting multiple access scheme, which is considered in this book. In such a system, the collisions between transmitted packets are acceptable as long as the level of multiple access interference is small compared to the strength of the power of the desired packet.

The geographical traffic nonuniformity problem is basically not the case for the conventional geostationary satellite systems, because of relatively wide coverage of a single geostationary satellite to about one-third of the globe. However, for a LEO satellite system, in which the coverage of a single satellite can be as small as a part of a country or an ocean, the problem becomes important. Generally, LEO satellite systems are planned to service all parts of the globe, including areas with relatively small numbers of users. In addition, in urban areas, the number of the future hand-held PCN terminals with the dual capability of direct access to the satellite system and their source country cellular system is expected to be large. The service area of a LEO satellite may cover a number of such small cities as well as the urban areas. Then the total traffic of the satellite becomes much higher than that of its neighbor satellite. In short, this problem results in nonoptimal usage of the communication facilities of the LEO satellite systems.

This book presents analytical frameworks for evaluating the performance of the LEO satellite systems under those specifications. A number of techniques to improve the performance of those systems are introduced. Those techniques are grouped into two types. The first group includes methods that are modified versions of the conventional power control necessary in spread spectrum systems. In such methods, according to the average level of traffic loads of satellites, different required users' transmitting powers are requested. Different types of these methods are employed in both CDMA and spread-slotted Aloha systems. By numerical examples, it is shown that they can improve significantly the signal-to-interference ratio and throughput characteristics of the LEO satellite systems. As will be shown, these methods are proper solutions to the nonuniform traffic distribution problem. The second group considers the control of transmissions of users to achieve significant improvement in the performance of the LEO satellite systems in both uniform and nonuniform traffic distributions. The method of controlling the transmissions of users enhances the throughput characteristics of the LEO satellite system comparably higher than those that can be achieved in a conventional spread-slotted Aloha scheme. It also maintains the improved characteristics in a wide range of change of the offered traffic load.

Organization of this book

Chapter 1 discusses the general ideas of applying satellites in communications systems. It also briefly describes the orbital dynamics in satellite systems. An overview of conventional geostationary satellite systems is followed by some objections to those systems, such as the need of low elevation angles at high latitudes as well as large propagation loss and delay. After that, we present some proposals for the LEO satellite systems that are evidence for the necessity of consideration of LEO satellite systems in future mobile communications.

Chapter 2 introduces the concept of communications with LEO satellites. The chapter presents some preliminary issues in those systems, including the calculations of the required number of satellites and orbits in a global satellite constellation, the concept of hand-off between LEO satellites for a continuous communication, the issue of networking

LEO satellites via intersatellite links, the idea of spot-beam antennas, and the problem of Doppler shift. After that, the text discusses two specific issues in a LEO satellite system: the selection of multiple access and the problem of traffic nonuniformity. The chapter discusses the meaning of and alternatives to the multiple access schemes in general and in LEO satellite systems specifically. The chapter finishes by introducing the mathematical model of a LEO satellite system and its alternative, which will be used throughout the rest of the book.

Chapter 3 examines application of CDMA in LEO satellite systems. The chapter focuses the discussion on an analog system and derives the signal-to-interference ratio as the measure of the performance in such a system. We introduce a mathematical nonuniform traffic distribution model and compare the performance of the system under uniform and nonuniform traffic distributions. After that, the discussion continues in an integrated voice/data scenario. In both cases, we propose a control scheme on the level of the transmitting power of the users and show the effect of such control on the performance of the system.

Chapter 4 introduces the combination of spread spectrum and slotted Aloha multiple access schemes. A spread-slotted Aloha scheme is intro-duced and then such a composite multiple access is applied on the uplinks of the LEO satellite communication system. The chapter explains the conventional (unspread) Aloha schemes as well as the combination of them with CDMA. After that, we present the necessary mathematics for the calculation of the throughput in the LEO satellite systems employing spread-slotted Aloha. We also compare the performance of the system under uniform and nonuniform traffic distributions and show how the traffic nonuniformity degrades the average value of total throughput in the system.

Chapter 5 proposes a new method for improving the throughput performance of LEO satellite systems by searching the worst case of the performance of the system. The chapter proposes a modified power control scheme applicable in a spread-slotted Aloha LEO satellite system faced with nonuniform traffic distribution. An analysis of the perform-ance of a LEO satellite system in different traffic situations is also presented, and some practical considerations for applying this method in a real satellite system are also examined.

Chapter 6 introduces the concepts of controlling the transmissions of users, namely, a transmit permission control scheme. The chapter provides the mathematical calculations of the average delay in both uniform and nonuniform traffic distributions. The performance improvement of the proposed scheme is shown in fading and nonfading satellite channels. It also is shown that the proposed method can be applied in both uniform and nonuniform traffic situations. After that, a modification to the proposed method based on an adaptive control of the transmissions to improve the performance more is presented.

The last chapter discusses some further considerations of LEO satellite systems. The chapter proposes a packet admission control scheme that is very similar to the transmit permission control scheme. In the new scheme, which is again applicable in a spread-slotted Aloha system, transmission of packets is controlled according to the distance of users to their connecting satellites as well as traffic distribution. It is shown that the method can provide improved throughput performance in heavy traffic situations. Chapter 7 also examines some imperfections in the system, especially the one that appears in power control. The effect of an imperfect power control and the sectorizations of antennas on the performance of the system and how they change the mathematical results given in other chapters are some of the subjects of this chapter. The chapter finishes by introducing concepts of adaptive array antennas, recently proposed for the LEO satellite systems.

1

Mobile Satellite Communications

THE EXPONENTIAL INCREASE in the number of subscribers for mobile telephones during the last five years can be assumed to be the trend of future mobile communications systems. Rather than the simple voice communications of the 1980s and the early 1990s, people now ask for a wide variety of personal communications, including voice, data, facsimile, and electronic mail, made available by the exploitation of wireless spectrum and the development of low-cost, low-power communications devices. In different countries, such systems are referred to as PCNs, *personal communications services* (PCS), *universal mobile telecommunications services* (UMTS), *universal personal telecommunications* (UPT), and most recently, the *future public land mobile telecommunication system* (FPLMTS). Such systems and services proposed to reach their ultimate goal by providing reliable, ubiquitous, and cost-effective communications to personal subscribers, either universally or continentally.

In addition to a wide variety of services, consumers are now seeking a single terminal and a single access number that can be used internationally. Unfortunately, there are many different standards through the world. Each continent or even each country has its own standard, which requires a different terminal even for voice communications. One example of such an idea is now realized partly by the Japanese *personal handy phone system* (PHS, formerly PHP), in which a user can use a small hand-held terminal as a cellular mobile phone and as a cordless phone connected to a home telephone line. Although we still are a bit far from complete realization of such a single-terminal, single-number system, we should expect it in near future.

This book focuses on a strong candidate for realizing such a system: The LEO satellite system. This chapter briefly describes satellite communications systems in general; subsequent chapters examine satellites in low-altitude orbits.

1.1 Communications satellites

1.1.1 Preliminary issues

It was not until about four centuries ago that the realization was made that the shape of our planet is spherical. As a direct consequence of that shape, it is impossible to send radio waves directly from one point on the globe to another point when the receiver point is not in the line of sight of the transmitter. Hence, a middle point must receive the signal and transmit it to the next visible point until the path between the original transmitter and the final receiver is complete. The middle points can be, for example, relay stations with tall antenna towers, as shown in Figure 1.1(a). However, since so many parts of the globe are occupied by water, it is impossible or very expensive to use such towers. That kind of relay station can be used for communications between far points only on land.

To establish long-distance communications between continents, another possibility is to use the Earth's atmosphere or the ionosphere layer.

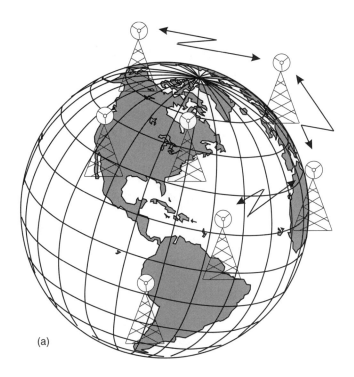

(a)

Figure 1.1 Different methods for communiactions between two locations on the Earth: (a) the use of tall antenna towers on land masses of the Earth; (b) the use of the Earth's atmosphere as a natural reflector; and (c) the use of satellites as man-made reflectors in the sky.

If radio signals are sent toward that layer and reflected off it, at least an attenuated form of the original signal can be received in another location on the Earth, as shown in Figure 1.1(b). Shortwave communications is an example of this method, in which the electromagnetic waves from a transmitter are bounced between the Earth's surface and the ionosphere to arrive at receivers. Limited bandwidth is one important problem with this method. Another problem is that the Earth's atmospheric conditions and its attenuation factor change often, depending on many uncontrollable parameters. If we think of the atmosphere as a simple reflector of electromagnetic signals, then other natural objects in the space, such as the moon, the planets, and stars, could also reflect signals.

Another alternative, derived from the reflection method, is to establish some artificial stations in the space that can receive radio signals and

Figure 1.1 (continued).

transmit them to another point on the Earth—a ground station, a relay antenna, or the final destination receiver, as shown in Figure 1.1(c). This is the basic idea of man-made satellite communications systems, used for many years until now.

Thus, we can define a communications satellite as a means for communication between two widely separate points on the ground. Although that definition seems simple, it is not well known. Many people think of a satellite as a means for broadcasting television signals. Think of the many homes equipped with satellite dishes used for television and of weather photographs taken from satellites and shown on the news. Here, however, we are defining a satellite as an essential part of global telecommunications carrying large amounts of data and telephone traffic world-

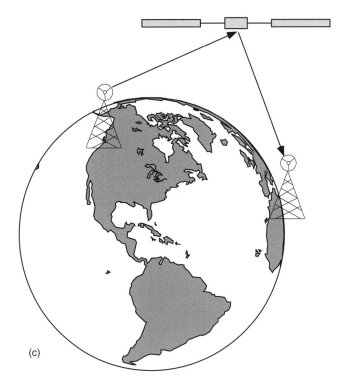

Figure 1.1 (continued).

wide. A telecommunications satellite also can be thought of as a star point in the sky receiving data from one point and transmitting them to several points on the ground.

We should note that a communications satellite can do many activities other than simple reflection of radio signals, such as switching facilities, navigation (e.g., *global positioning system* (GPS)), information processing, and remote sensing. Such activities are determined according to the payload of the satellite and the purpose for which the satellite is launched. Throughout this book, the word *satellite* is used indicate telecommunications purposes; we do not discuss, for example, broadcasting satellites.

1.1.2 History of communications satellites

With the advent of man-made satellites, extensive research and development work has made in various countries to utilize the satellites as a means of long-distance telecommunications. The result was rapid progress in satellite communications systems. Today, satellite communications are indispensable as a basic tool of human social activities. This system, as an epoch-making, modern communication means, is now broadly utilized not only in telecommunications but also in broadcasting, meteorological observations, navigation, and resource exploitation as well as space research.

A communications satellite provides a number of features not readily available with other means of communications. Perhaps the most important feature of a satellite is its unique ability to cover wide areas on the Earth's surface. As a consequence of that wide coverage, a satellite can form the star point of a communication network linking many users simultaneously, users who may be widely separated geographically. Moreover, the wide coverage of a satellite enables communication in sparsely populated areas that are difficult to access by other communication means. It is worth mentioning that providing communications between small cities located great distances apart is an expensive task if we ignore the satellite as a means of communications.

As already mentioned, a satellite can be used as a means for communication between two locations on the Earth separated by a large distance. If we consider the reflection role of a satellite in such communications, that is, receiving a signal from the source location and forwarding it to the destination location, and if we agree that such communication is repeated at different hours every day, then maybe the most proper reflector will be the one that is fixed from the viewpoint of an object on the Earth. Because the Earth is continuously rotating, the satellite should also rotate with the same angular speed and in the same direction as the Earth, in order to be fixed with any objects on the Earth. That is the concept behind launching satellites on the *geostationary Earth orbit* (GEO). A satellite on a GEO is referred to as *geostationary satellite* or, in some literature, as a GSO (*geostationary satellite orbit*) satellite.

As will be discussed in Section 1.2, it can be shown by mathematical analysis that there is only one GEO and that it is at an altitude of about 36,000 km and in the equatorial plane. When the position of a satellite is always stationary related to the Earth, the synchronization process between satellite and Earth stations becomes simple. In addition, with three geostationary satellites rotating in the plane of the equator, separated by 120 degrees of longitude, it is possible to cover almost all parts of the land masses on the Earth, except for the north and the south polar regions. Simplicity in synchronization in addition to global coverage by only three satellites were why satellite systems on geostationary orbit were so successful in last three decades.

The most noteworthy achievement in satellite communications is that in 1964 the *International Telecommunications Satellite Organization* (INTEL-SAT) was established to provide a means of fixed-satellite service among nations and that as early as 1965 satellite communications were put into practical commercial use. The stage of development up to the practical application of satellite communications, however, would be the age of experimental space radio communication, detailed descriptions of which are available in much of the literature [1–6]. The *International Maritime Telecommunication Satellite Organization* (INMARSAT), another key-pioneered satellite system for mobile purposes, is discussed in Section 1.3.

1.2 Orbital dynamics of satellite systems

Before discussing our main topic, that is, communications with LEO satellites, we should review the dynamics of satellite systems. Because this book is from a communications engineering viewpoint, we will not discuss either the dynamics of the orbits or their mechanics in detail. For those subjects, the reader is referred to well-written books on the dynamics of satellite systems, for example, Roddy; Elbert; and Pritchard, Suyderhoud, and Nelson [1,4,5].

A satellite is an artificial body in space, but it has to follow the same laws in its rotation as the planets do in their rotation around the sun. Three important laws for planetary motion derived empirically by Johannes Kepler (1571–1630) were derived again by Isaac Newton, in 1665, according to Newton's laws of mechanics and gravitation theory. Kepler's laws are general and can be applied to any two objects in space. It is usual to refer to the more massive object as primary and the smaller one as secondary. Using those labels, for a satellite rotating around the Earth, the Earth is the primary object and the satellite the secondary object. The following explanations of Kepler's three laws can be used to describe satellite systems as well. We use the words *Earth* and *satellite* instead of *primary* and *secondary*, respectively, to emphasize the application of Kepler's laws to satellite systems.

1.2.1 Kepler's first law

Kepler's first law states that when a satellite rotates around the Earth, its rotating path is on an ellipse, with the Earth on one of the two focal points of that ellipse. If we denote the semimajor axis and the semiminor axis of the ellipse by r_a and r_p, respectively (Figure 1.2), then the eccentricity parameter, e, can be defined as

$$e = \frac{\sqrt{r_a^2 - r_p^2}}{r_a} \tag{1.1}$$

The semimajor axis and the eccentricity are the two orbital parameters in satellite communications systems. Note that in the case of $e = 0$, the orbit becomes circular. The point in the orbit where the satellite is closest to the Earth is called the *perigee*, and the point where the satellite is farthest from the Earth is called the *apogee*. Therefore, the semimajor and semiminor axes sometimes are referred to as the apogee radius and the perigee radius, respectively.

1.2.2 Kepler's second law

Kepler's second law states that in equal time intervals, a satellite will sweep out equal areas in its orbital plane. For example, Figure 1.2 shows that the satellite sweeps out the equal areas indicated by $a1$ and $a2$. If we denote the average velocity of the satellite during its sweeping of areas $a1$ and $a2$ by $V1$ (m/sec) and $V2$ (m/sec), respectively, it is obvious that $V2 < V1$. Using this law, we will show later that a GEO should be circular, not elliptical. Kepler's second law also states that if a satellite is far from the Earth, there is a longer time during which the satellite is visible from the viewpoint of a specific object on the Earth.

1.2.3 Kepler's third law

Different from the first and second laws, Kepler's third law provides more mathematical facilities. Kepler's third law states that there is a

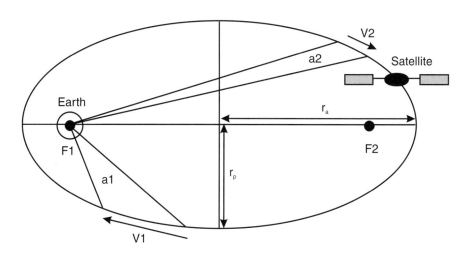

Figure 1.2 An illustration of the orbit parameters used in a satellite system.

relation between the periodic time of orbit, that is, the time required for one complete orbit, denoted by P_0, and the mean distance between satellite and the Earth. The mean distance between the Earth and the satellite is equal to the semimajor axis, r_a; then, the third law can be shown in the form of an equation as

$$r_a = AP_0^{2/3} \tag{1.2}$$

where A is a constant, which can be determined according to the dimensions of r_a and P_0. With r_a in kilometers and P_0 in mean solar days (a unit equal to 1.0027379 sidereal days that we use), the constant A for the Earth evaluates to $42,241.0979$.

It is worthwhile to show the other form of Kepler's third law, which was derived by Newton. That law of Newton finds the angular velocity of a satellite at any altitude very simply. According to this law of Newton, the angular velocity, ω_{vs}, of a satellite at the altitude h can be found from

$$\omega_{vs} = (gm)^{1/2} \cdot r^{-3/2} \tag{1.3}$$

where $(gm)^{1/2} = 631.3482 \text{ km}^{3/2}/\text{s}$; g is the gravity constant; m is the mass of the Earth; and r is the radius of the satellite orbit, equal to the sum of average equatorial radius of the Earth, R, and the altitude of satellite, h.

Because Kepler's third law provides a fixed relation between the period and the size, it can be used to find, for example, the rotation period of a satellite that is on a geostationary orbit. It should be noted that (1.2) assumes an ideal situation, one in which the Earth has a perfectly spherical shape and uniform mass. That equation also assumes that no perturbing forces, such as gravitational forces of the sun and the moon and atmospheric drag, are acting on the orbit. The gravitational pulls of the sun and the moon have a negligible effect on LEO satellites, but they do affect satellites in geostationary orbit. On the other hand, atmospheric drag affects mostly satellites on lower orbits and has negligible effect on GEO satellites.

1.2.4 An example: The geostationary orbit

For an example of an application of Kepler's laws, consider the evaluation of altitude of the geostationary orbit. We will show that there is only one orbit in the equatorial plane on which a satellite can rotate around the Earth in a 24-hour period, and that altitude is about 35,780 km.

As mentioned before, a geostationary orbit is the orbit on which a satellite appears stationary relative to any objects on the Earth. When a satellite is on the geostationary orbit, the antennas of ground stations can be kept pointed to the satellite automatically, because the Earth is rotating with the same period as the satellite. That makes the tracking process for antennas simple.

For a satellite to be stationary with the rotation of the Earth, it is not enough only to have a geosynchronous orbit, that is, one that has the same orbital period as the Earth's spin period. A satellite on any geosynchronous orbit with some inclination other than zero would appear to move in a figure-eight pattern when viewed from a fixed location on the Earth [1]. (The inclination angle is the angle at which a satellite orbit is tilted relative to the Earth's equator. That is, it is the angle between the orbital plane and the Earth's equatorial plane.) On the other hand, to have the constant angular velocity for a satellite the same as that of the Earth, Kepler's second law requires a circular orbit. Therefore, a geostationary orbit is only a circular orbit in the equatorial plane, that is, with zero inclination, and has the same orbital period as the Earth.

To find the altitude of the geostationary orbit, we can use Kepler's third law. If we denote the altitude of the satellite and the average equatorial radius of the Earth by h and R, respectively, then for the circular orbit, we have

$$r_a = r_p = R + h \qquad (1.4)$$

It can be shown that [1] for the geostationary orbit P_0 defined in (1.2) is equal to 0.9972695. Then, according to the Kepler's third law, we have

$$R + h = 42241 \cdot (0.99727)^{2/3} \qquad (1.5)$$

which, with $h = 6378.14$ km, results in an altitude of 35,786 km for the geostationary orbit. Because (1.4) has only one numerical answer, we can say that there is only one geostationary orbit for the Earth that is in the equatorial plane. Any other orbit at some inclination other than zero could not to be referred to as a geostationary orbit.

The fact of having only one geostationary orbit emphasizes that it should be used efficiently. As for any two successive satellites on GEO, there should be enough spacing to avoid physical collisions between satellites, there is a limitation on the number of geostationary satellites. Currently, there are hundreds of geostationary satellites that belong to different countries. The available frequency spectrum assigned to GEO satellite systems is a more important limitation for these systems. The two limitations imposed by the problems of frequency spectrum utilization and space utilization can be considered as reasons for launching satellites to orbits other than the geostationary orbit.

1.3 Mobile satellite communications systems

1.3.1 Orbit selection

1.3.1.1 Problems with geostationary satellites

Much research has been dedicated to establishing a common, global standardization for communications. Satellites are the only means of providing coverage to all parts of the globe, even those parts for which the communications service is a very expensive or difficult task. There is always a question on the best Earth orbit constellation that can realize an appropriate global communications service [7]. Unfortunately, satellites in geostationary orbit could not support all the requirements for future global communications systems, perhaps chief among them being the size of terminal required in the next generation of communications systems. A satellite in geostationary orbit has many advantages, such as wide coverage, high-quality and wideband communications, availability for mobile communications, and economic efficiency. Also, their synchronization with the rotation of the Earth makes the tracking process much simpler than the one required for nongeostationary orbits. However, GEO satellites suffer from some disadvantages when compared to other lower-altitude orbits.

A satellite in the geostationary orbit suffers from long propagation delay, which is completely unavoidable because of the great distance from the Earth and the finite velocity of electromagnetic waves. As discussed in Section 1.2, a geostationary satellite has an altitude of about 35,780 km. Considering the velocity of light, $3 \cdot 10^5$ km/s, a two-way propagation delay, including the uplink and the downlink, is between 240 and 270 ms, depending on the elevation angle from the position of a user to the satellite, as shown in Figure 1.3. A typical international telephone call requires a round-trip delay on the order of 540 ms. In a voice communication system, such a delay can cause echo effect during conversations, which can be repaired by echo-suppresser circuits. However, in the case of data communications, that delay makes errors in data, so error-correction techniques are required.

Another disadvantage of a satellite on geostationary orbit similar to the long propagation delay is its large propagation loss. In a satellite communication system, the power of electromagnetic signals is attenu-

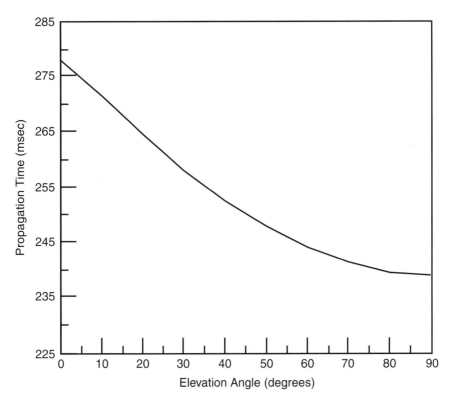

Figure 1.3 Relationship between elevation angle and propagation delay in a geostationary satellite system.

ated with the second power of the distance that the signal propagates. For example, if the propagation distance between a transmitter and a receiver becomes double, we need four times the power level at the transmitter to have the same power level at the receiver. If we think about future hand-held mobile terminals with limited power supply, that high-power requirement will not allow use of a satellite on the geostationary orbit. Even with the current high technologies of batteries and hardware, the smallest terminal for a geostationary satellite is as large as the size of an A4 paper and as heavy as 2.5 kg (used in standard mini-M of INMARSAT-M).

The next fundamental objection to a geostationary satellite is the lack of coverage at far northern and southern latitudes. Because a geostationary satellite is flying in the plane of the equator, many areas with high latitudes require very low angles of elevation to access the satellite. However, experimental measurements have shown that for consistent service, especially in urban areas, elevation angles as high as 40 degrees are desirable. Such high elevation angles are difficult to achieve with geostationary satellites even in the capitals of Europe. As we will discuss later, with polar low Earth orbital constellation, those high elevation angles are easily achievable.

These objections to geostationary satellites, along with other problems, such as the high cost of launching a satellite into geostationary orbit and the influence on the space station of an eclipse, suggest the use of other orbits for mobile satellite communication systems. Especially, it is possible to have short propagation time and loss (i.e., smaller-size users' terminals), as well as high elevation angles at high latitudes by the constellation of satellites on LEO or *medium Earth orbit* (MEO). Although we have only one geostationary orbit and limited space for a constellation of satellites, there are (at least theoretically) an infinite number of nongeostationary orbits. That gives the satellite system designer much more flexibility in network architecture.

1.3.1.2 Comparison of different orbits

Even though it may seem that the altitude of a satellite can be freely chosen, the existence of two Van Allen radiation belts limits orbit selection. As illustrated in Figure 1.4, the two Van Allen belts are centered on the Earth's geomagnetic axis, at altitudes ranging from 1,500 to 5,000 km and from 13,000 to 20,000 km. To minimize the radiation damage to electronic components that would result from a relatively unshielded, lightweight satellite, as in the case of LEO satellites, it is better to put the satellites out of these belts. Extensive ionizing radiation severely reduces useful satellite life. Many LEO or MEO satellite system proposals consider the altitude outside these two belts, as are shown in the figure.

Although serious consideration of LEO satellite systems for commercial purpose did not start until the 1990s, even in the early 1960s there was a comparison study of the merits of GEO versus LEO and MEO [8]. In that study, the convenience of GEO was weighed against the practical difficulty of attaining it and the inherent technical advantages of LEO, such as less time delay and higher angles of elevation. While it was

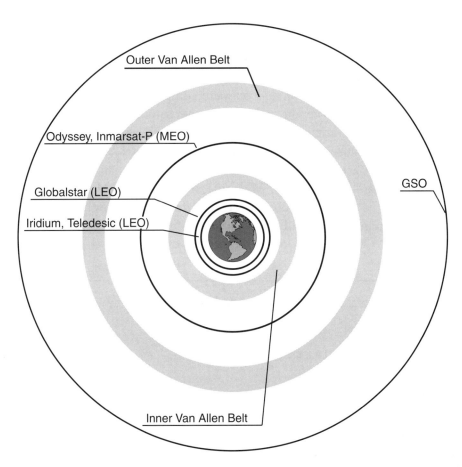

Figure 1.4 Orbit altitude selection for satellite systems.

conceded that GEO was in many respects theoretically preferable, the state of technology at the time suggested that LEO or MEO systems were preferred in the near term. The orbit selection in satellite systems has taken the attention of many researchers for a long time [9–13].

This subsection briefly presents a comparison of different orbit constellations. According to Kepler's laws, we can divide the orbit of satellites into two groups: Circular and noncircular (elliptical). Another categorization can be made according to the altitude of the orbits, which communications engineers often use. According to the latter categorization, we have GEO at an altitude of 35,786 km; MEO at an altitude of 10,000 to 20,000 km, and LEO at altitudes less than 1,500 km. This book is concerned with circular orbit satellite systems; hence, we will not discuss *highly elliptical orbits* (HEO), for example proposed in ELLIPSO system of Ellipsat.

Figure 1.5 illustrates an approximate comparison of the number of satellites for global coverage, relative cost per satellite, and relative cost for launching different proposed satellite system constellations. As it can be seen from Figure 1.5, as the altitude of the satellites becomes lower, more satellites are required for global coverage. For example, the proposed LEO satellite system by Motorola, named IRIDIUM, requires 66 satellites for its complete global coverage plan. On the other hand a GEO satellite system requires only three satellites to cover the Earth.

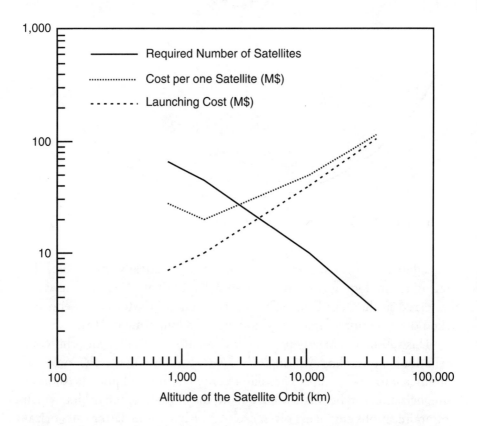

Figure 1.5 Comparison of satellite systems according to their altitudes.

In the case of both the launching cost and the manufacturing cost per satellite, as shown in Figure 1.5, the GEO satellites are the most expensive systems. However, when we consider the number of satellites in each system, a LEO satellite system is much more expensive. Table 1.1 compares the three constellations of LEO, MEO, and GEO satellite systems. As the table shows, the most expensive and the most complicated system is the one whose satellites are in LEOs. In that case, the satellites are rotating rapidly in their orbits; hence, the synchronization process requires complex facilities, which is almost unnecessary in the case of GEO satellite systems. On the other hand, the small coverage area of a single LEO satellite dictates a large number of satellites for global coverage. That is why LEO satellite systems sometimes are referred to as networks in space. However, because only LEO satellite systems offer the advantages of low propagation delay and loss compared to other systems, that makes them candidates for a future global personal mobile communications network. Figure 1.6 is a simple view of a future LEO satellite communications system, in which the satellite system has close cooperation with the current terrestrial mobile systems and the public telephony networks.

Table 1.1
Comparison of Different Satellite Systems

	LEO	MEO	GEO
Satellite cost	Maximum	Minimum	Medium
Satellite life (years)	3–7	10–15	10–15
Hand-held terminal	Possible	Possible	Very Difficult
Propagation delay	Short	Medium	Large
Propagation loss	Low	Medium	High
Network complexity	Complex	Medium	Simple
Hand-off	Very	Medium	No
Development period	Long	Short	Long
Visibility of a satellite	Short	Medium	Always

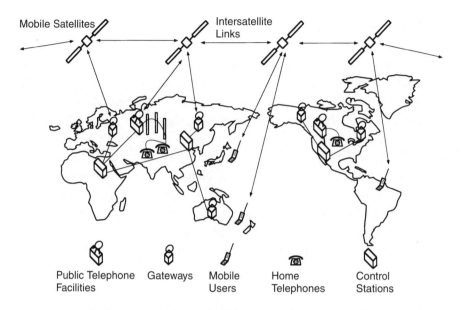

Figure 1.6 Example of a future personal mobile communications system provided by LEO satellites.

1.3.2 Mobile satellite systems

More than three decades after the first operational communications satellite was launched, we finally are seeing a rush of applications of satellites for mobile communications. In past years, the satellite was the only means for transferring telecommunications data and television broadcasting signals over long distances. In more recent years, however, due to the growth of fiber-optic cable technology, many countries prefer to use fiber-optic cables for their international telephony systems. Currently, there are some international projects to establish long-distance telecommunications between continents with under-ocean fiber-optic cables. Many countries, such as Japan, are now using fiber-optic cables for their long-distance national telecommunications. A fiber-optic cable can provide voice and data transmission that is much cheaper, of much better quality, and more reliable than that provided by satellite channels. One cable can include thousands of communication channels; after a fiber-optic communications system has been established, the cost per call can be reduced easily.

Although fiber-optic cables can provide cheap telecommunications for high-traffic, fixed applications worldwide, for mobile communications they would not be useful. Moreover, even for some fixed applications, establishment of fiber-optic systems in rural areas with low communication traffic, such as in small towns, is not economical. For mobile communications as well as communications in isolated areas, wireless systems, especially satellite wireless systems, will be around for a long time.

Mobile communications with satellites can be put into two different categories according to the satellite constellations. In communications with GEO satellites, because the satellites are synchronized with the Earth, mobile communication includes the communication of physically mobile objects through satellites. However, with satellites that are not in synchronization with the Earth, such as satellites in nongeostationary orbits, mobile communications can include both fixed and mobile objects on the Earth through those satellites. Note that with a typical value of $h = 1,500$ km in a LEO satellite system, the satellite has a linear ground speed of about 7.1 km/s, not comparable to the speed of any object of the Earth. Therefore, any mobile or fixed terminal on the Earth can be assumed to be a fixed object in a LEO satellite system.

The anticipated global cellular and paging market size through the next century will mandate satellites for mobile communication. It is anticipated that by the year 2000 the number of global mobile phones and pagers will be more than 140 million each. Satellite systems are the only means for supporting such a large number of terminals. LEO satellites, especially, will play a major role in preparing the communications industry for the 21st century.

The following subsections describe some mobile satellite communications systems currently under operation. These systems use GEO satellites; hence, their mobile terminals still are too large and too heavy to be considered for future global PCNs. We also examine some of the proposals that use satellites in lower orbits. Generally, LEO satellite systems have been categorized into two groups: "big-LEO" and "little-LEO," according to the payload of the satellites, their frequency bands, and the variety of services they provide. Simply said, big-LEO systems have larger capacity and voice transmission, whereas little-LEO systems have less capacity and limited nonvoice services. Because LEO satellite systems are still under development, interested readers can get more updated information on each system in publications from the satellite companies. The information provided in this chapter is given only to introduce the high-speed progress in LEO satellite communications systems, commercially started just in the current decade. It is noteworthy that in the less than 10 years since 1990, in the United States alone, 17 companies have filed applications or announced plans to develop satellite systems in nongeostationary orbits. Some analysts project that by the year 2000 there will be 813 small satellites in nongeostationary orbits, increasing to 1,322 satellites by the year 2005. A detailed discussion of the regulatory issues for the new age of satellite systems can be found in [14].

1.3.2.1 Mobile satellite systems in operation

The beginning of the first generation of mobile satellite communications systems (sometimes referred to as *mobile satellite services*, or MSS) was signaled by the start of service of INMARSAT, in 1982. INMARSAT stands for *INternational MARitime telecommunication SATellite* organization and is a GEO satellite system using *L*-band (1.5–1.6 GHz) mostly to provide different services to ships. In the first generation of MSS, INMARSAT defined five standards: standard A (1982), standard B (1993), standard C (1991), standard M (1992/1993), and aeronautical standard (1992). All these standards considered different services worldwide, including voice, facsimile, and data. INMARSAT A and B are mostly considered the service to ships, and INMARSAT C is planned to service to small crafts, fishing boats, and land mobiles. The standard aeronautical service is a bit different from others because it considers the service to commercial and private aircraft. The weight of INMARSAT terminals in its different standards ranges between 25 kg in standard A to the lightest one in standard C, about 5 kg. In 1996, there were about 71,900 INMARSAT terminals around the world; more than one-third of them maritime terminals using standard A.

For the first generation of mobile satellite communications, we also can consider other systems such as QUALCOMM, started in 1989 and servicing North America; ALCATEL QUALCOMM, in 1991 for Europe; and the Japanese system NASDA, in 1987 for their national services.

Around 1995, the second generation of MSS was started to reduce the size and the cost of user terminals, and also internetworking with terrestrial systems. In this generation, INMARSAT defined its mini-M standard in 1995 with worldwide voice, data, facsimile, and telex services at a 2.4-Kbps data rate. This standard is realized by its small terminals, laptop size and weighing 2.5 kg. In 1996, mini-M had 150 users. *American Mobile Satellite Corporation* (AMSC), NSTAR of Japan, *European mobile satellite* (EMS), and OPTUS of Australia are other satellite systems included in the second generation of MSS.

Detailed descriptions of those GEO-based mobile satellite communication systems, with either global or continental service, can be found in some literature [9,10]. The two-decade age of these GEO satellite systems shows that, although their services are acceptable for maritime and aeronautical purposes, they are not good candidates for future personal communication systems. Future personal communication systems will require very small, light users' terminals, similar to the ones now used in cellular systems, which is not expected to be realizable in those GEO systems. The need for such small terminals should be found through nongeostationary orbits, especially the LEOs.

1.3.2.2 Little-LEO mobile satellite systems

Little-LEO mobile satellite systems are a category of LEO systems that utilize satellites of small size and low mass for low-bit-rate applications under 1 Kbps. The *Federal Communications Commission* (FCC) has allocated frequency bands of 137–138 MHz for downlinks and 148–149.9 MHz for uplinks to these systems, which is a heavily utilized spectrum worldwide for private and government services. Now three organizations have received their licenses, namely, *Orbital Communications Corporation* (ORBCOMM), with 36 satellites at the altitude of 775 km; *Volunteers in Technical Assistance* (VITASAT); and STARNET, with 24 satellites at the altitude of 1,000 km. Several others have proposed systems. ORBCOMM has a national service plan in the United States, and both VITASAT and STARNET have global services. The mass of satellites in these systems ranges from 40 kg in ORBCOMM to 150 kg in STARNET. Nonvoice two-way messaging and positioning with low-cost transceivers, which would be equipped with alphanumeric displays, are the major characteristics of these systems.

Little-LEO satellite systems prefer a spectrum below 1 GHz, because it enables the use of lower cost transceivers; however, that spectrum is heavily utilized worldwide. Little-LEO proponents are continuing to work with regulatory offices to identify suitable frequency bands for future use, but it is not likely that such frequencies would be available on a global basis.

1.3.2.3 Big-LEO mobile satellite systems

Another category for LEO satellite systems is the so-called big-LEO satellite systems. Compared to the little-LEO systems, satellites in big-LEO systems are expected to be bigger and to have more power and bandwidth to provide different services to their subscribers. Those services may include voice, data, facsimile, and *radio determination satellite services* (RDSS). Big-LEO systems will use the underutilized spectrum available in the *L*-band, because of the commercial failure of the proposed RDSS systems. Currently, the frequency spectra of 1610–1626.5 MHz for uplinks and 2483.5–2500 MHz for downlinks are assigned to these systems. It is interesting to note that, although the names of these systems include LEO, their frequencies are the ones usually utilized in MEO and GEO satellite systems.

Most of the proposed big-LEO systems would offer global service to hand-held terminals by the means of satellites on lower altitudes moving very fast instead of a fixed point in the sky, as for GEO satellite systems. The bigger size of the satellites in these systems enables them to have more complex data-processing facilities in the satellite than the simple store-and-forward feature of satellites in little-LEO systems. An important parameter in the development of these systems is their internetworking with terrestrial systems and, especially, dual-mode terminals. That makes the system more economical in different parts of the globe and also more flexible to handle increasing traffic loads in future.

Several big-LEO systems are being proposed. Some of the more important systems are described here. Interested readers are referred to more detailed descriptions of these systems [15–19].

IRIDIUM Motorola proposed its big-LEO satellite system, IRIDIUM, for global coverage and a variety of services, including voice (full-duplex,

2.4 Kbps), data (2400 baud), facsimile (2400 baud), paging, and RDSS [18, 19]. Sixty-six satellites in the IRIDIUM system are at the altitude of 780 km on six polar orbit planes. (Specifically, the orbital planes of the IRIDIUM system are near-polar with inclination of 86.4 degrees.) Each IRIDIUM satellite has three *L*-band antennas, which project 48 spot beams onto the Earth, to form 48 cells at the footprint of each satellite, totaling 3,168 cells, of which only 2,150 need to be active to cover the entire surface of the Earth; in other words, it is a cellular-type satellite system. Each IRIDIUM satellite has a mass of about 700 kg, with a lifetime of 5 to 8 years. The satellites of this system have complete information transference by utilization of intersatellite links, which make the system a network in the sky. Figure 1.7 shows a schematic illustration of the IRIDIUM system; Figure 1.8 shows the satellite and orbit constellation of the system and its intersatellite link view.

The IRIDIUM system is proposed to be in complete cooperation with the existing terrestrial system. The dual-mode hand-held transceivers of IRIDIUM would first try to access local cellular telephones before using the satellite system. If it is not possible to use the terrestrial systems, because of long distance or overload traffic on those systems, the terminal would automatically switch to its satellite mode. Motorola has proposed bidirectional operation in the *L*-band (1616–1626.5 MHz); that is, the same frequencies would be used for uplinks and downlinks on a time-shared basis. Messages from one telephone to another would be transmit-

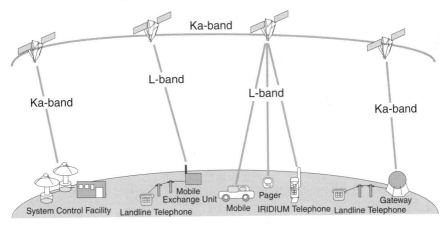

Figure 1.7   The features of the IRIDIUM system.

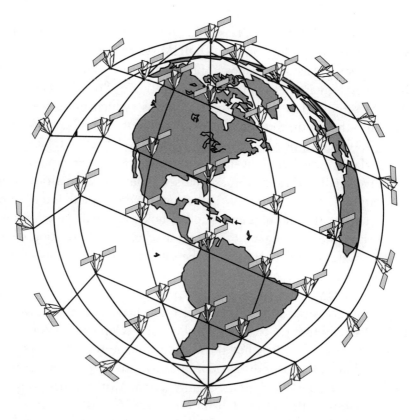

Figure 1.8   Satellite and orbit constellation for the IRIDIUM system and the intersatellite links in that system.

ted from the hand-held unit to the satellite and then transmitted from satellite to satellite using *Ka*-band (23.18–23.38 GHz) intersatellite links until the satellite viewing the destination telephone is reached. This system also has gateway/telemetry, tracking, and control links working at 19.4–19.6 GHz (downlinks) and 29.1–29.3 GHz (uplinks) again on *Ka*-band, as shown in Figure 1.7. IRIDIUM satellites have on-board processing facilities that can demodulate a signal, read addresses, and route signals directly to one of the four adjacent satellites: two in the same orbit (north-south) and one in each adjacent (east-west) orbit via intersatellite crosslinks.

The IRIDIUM system uses *time division multiple access* (TDMA) as the multiple access scheme and *time division multiplexing* (TDM). The connections to the terrestrial telephone network would be via gateway Earth stations that could be regional or even in each country, as shown in Figure 1.7. IRIDIUM is the only big-LEO system that has on-board processing with switching via cross-links. It also is the only system that uses TDM, so only one band (1616.0–1626.5) is used for both uplink and downlink communications. The major problem in development of this system would be the difficulty in synchronization of its TDMA frames during the fast motion of satellites at the altitude of 780 km.

GLOBALSTAR   Another strong proposal for a big-LEO system is that from Loral Qualcomm Satellite Services. GLOBALSTAR is different from IRIDIUM in that it uses a *code division multiple access* (CDMA) scheme. GLOBALSTAR would provide global voice, data, facsimile, and RDSS services via its 48 satellites in eight inclined orbits 1,414 km above the Earth. The footprint of each satellite in this system is divided into 16 cells by six spot-beam antennas to receive and send messages to hand-held terminals or to gateways. Again, the system is intended to work with the existing *public switched telephone network* (PSTN). Calls would be relayed through the satellite only when access could not be made to the terrestrial network. The existing PSTN would be accessed via gateways and used for long-distance connections, including transoceanic calls.

Unlike the polar-orbit inclination of IRIDIUM, GLOBALSTAR has a 52 degree orbit inclination. This system, again different from IRIDIUM, uses the bent-pipe approach to route long-distance calls. A simple network architecture for bent-pipe architecture, used in many LEO systems, is shown in Figure 1.9. In this architecture, each satellite establishes a moving footprint that is in communication with a gateway. The individual ground users establish a traffic link to the satellite via a spot beam within the footprint. All such communications must go up/down to/from the gateway. For users in the terrestrial network or in another satellite footprint, terrestrial lines must be used to complete the circuit. The bent-pipe approach is simple and inexpensive to build, but it depends on the terrestrial infrastructure for the networking, and many gateways are necessary for coverage.

ODYSSEY   The third big-LEO satellite system that we will mention here, is the satellite system proposed by TRW and known as ODYSSEY.

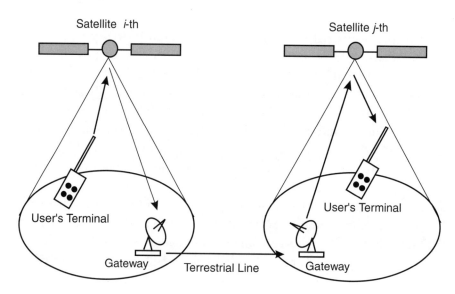

Figure 1.9 An example of the bent-pipe network architecture used in many LEO satellite systems.

ODYSSEY would provide global voice, data, facsimile, and RDSS services via its 12 satellites on three orbits of 10,354 km altitude. Though that altitude would be considered a MEO, according to frequency spectrum used by this system, it can be included in the big-LEO category. The ODYSSEY system would employ dynamically steerable satellite antennas to provide global coverage principally to land masses and the coastal areas of the world. As the satellites move over the Earth, the antenna beams would be steered to keep the satellite footprint stationary. The system is intended to work on a complementary basis with the existing terrestrial telephone systems, including cellular telephones. Orbit inclination of this system is 50 degrees. The multiple access scheme for this system is also CDMA with frequency division duplexing. ODYSSEY considers 37 spot beams in each satellite footprint.

ARIES Constellation Communications Inc. has proposed a LEO satellite system, named ARIES, that would provide global voice, data, facsimile, and RDSS. This system would consist of 48 satellites in four polar planes at an altitude of 1,020 km. Each ARIES satellite would have seven antenna beams to enable the system to provide global coverage. Hand-held terminal or vehicular transceiver units coupled to the satellites and gateway stations would provide interconnections to public or private terrestrial telephone networks.

1.3.2.4   Other mobile satellite systems

Several other LEO satellite systems have been proposed. One is the satellite system proposed by the *European Space Agency* (ESA) called LEONET. This system has 15 satellites on three orbits of 54 degrees inclination and 37 spot beams in each satellite footprint. The altitude of the satellites would be 6,390 km, that is, in the first Van Allen belt; hence, severe shielding would be required for each satellite's body.

INMARSAT proposed its *intermediate circular orbit* (ICO) system, which is a MEO satellite system, with 12 satellites on the 45 degree orbit inclination at an altitude of 10,400 km. ICO has been proposed with TDMA and *frequency division duplexing* (FDD) as multiple access and duplexing, respectively. This system also utilizes the bent-pipe architecture.

Another system, proposed by Ellipsat, is ELLIPSO. The system would provide voice, data, facsimile, and RDSS via its 15 satellites in elliptical inclined orbits and up to nine satellites in equatorial circular orbits with a maximum altitude of 7,800 km. The satellite system would provide coverage of the entire northern hemisphere and of the southern hemisphere up to 50 degrees south latitude. The elliptical orbit satellites would spend a greater proportion of their orbital periods over the northern latitudes. This orbital strategy would provide greater service capability to the areas having the greater land masses, the larger populations, and potentially the larger markets for services. ELLIPSO is also planned to cooperate with the terrestrial PSTN.

1.4   Summary

This chapter discussed the unavoidable role of satellites in long-distance communications. It briefly introduced the fundamental dynamics of satellite systems and discussed the advantages and disadvantages of a satellite on a geostationary orbit. It also examined the need for employing satellites on nongeostationary orbits in future personal communication systems.

The chapter reviewed the features of the most important mobile satellite communications systems for both geostationary and nongeostationary orbits. The systems described here, in addition to several others not mentioned, are evidence of the fact that the LEO satellite system is not a dream. Although in many features, such as synchronization with ground users and gateways, the LEO systems require much more complexity, their unique features of low propagation delay and loss and wide coverage area by each satellite are what future PCNs will require. Much research recently has been done on LEO satellite systems to show their important role in the future communications systems [20–37].

References

[1] Roddy, D., *Satellite Communications,* Englewood Cliffs, NJ: Prentice-Hall, 1989.

[2] Williamson, M., *The Communications Satellite,* Bristol and New York: Adam Hilger, 1990.

[3] Jansky, D. M., and M. C. Jeruchim, *Communication Satellites in the Geostationary Orbit,* Norwood, MA: Artech House, 1987.

[4] Elbert, B. R., *Introduction to Satellite Communication,* Norwood, MA: Artech House, 1987.

[5] Pritchard, W. L., H. G. Suyderhoud, and R. A. Nelson, *Satellite Communication Systems Engineering, 2nd ed,* Englewood Cliffs, NJ: Prentice-Hall, 1993.

[6] Miya, K., et al., *Satellite Communications Technology, 2nd ed,* Tokyo: KDD Engineering and Consulting, 1985.

[7] Adams, W. S., and L. Rider, "Circular Polar Constellations Providing Continuous Single or Multiple Coverage Above a Specified Latitude," *J. Astronaut. Sci.,* Vol. 35, 1987.

[8] Pritchard, W. L., "Criteria for the Choice of Synchronous or Medium-Altitude Systems," *IEEE Trans. Commun. Syst.,* Vol. CS-12, No. 2, 1964, pp. 131–137.

[9] *Global Mobile Communications Systems Under Development,* Washington, DC: Washington Business, Oct. 18, 1993.

[10] Wu, W. W., et al., "Mobile Satellite Communications," *Proc. IEEE,* Vol. 82, No. 9, 1994, pp. 1431–1448.

[11] Ananasso, F., and F. D. Priscoli, "The Role of Satellites in Personal Communication Services," *IEEE J. Select. Areas Commun.*, Vol. 13, No. 2, 1995, pp. 180–196.

[12] Vatalaro, F., et al., "Analysis of LEO, MEO, and GEO Global Mobile Satellite Systems in the Presence of Interference and Fading," *IEEE J. Select. Areas Commun.*, Vol. 13, No. 2, 1995, pp. 291–300.

[13] Katayama, M., and N. Morinaga, "Technical Feasibility of Communication Systems Using Non-Geostationary Satellites," *Proc. IEICE Joint Conference of Communication Network, Signal, and System,* 1990, pp. 211–215.

[14] Abeshhouse Stern, J., "Small Satellites: Worldwide Developments and Trends in Communications and Remote Sensing," *Tech. Report of IEICE,* SANE96-60, October 1996, pp. 23–49.

[15] M. Werner, M., et al., "Analysis of System Parameters for LEO/ICO-Satellite Communication Networks," *IEEE J. Select. Areas Commun.*, Vol. 13, No. 2, 1995, pp. 371–381.

[16] Pickholtz, R. L., "Communications by Means of Low Earth Orbiting Satellites," *Modern Radio Science 1996,* J. Hamelin, ed., Oxford Science Publications, U.R.S.I, pp. 133–149.

[17] Hatlelid, J. E., and D. E. Sterling, "A Survey of Small Spacecraft in Commercial Constellations," *Proc. AIAA/Utah State University Conf. Small Satellites,* Utah, Aug. 1991, pp. 2–7.

[18] Grubb, J. L., "The Traveler's Dream Come True," *IEEE Commun. Mag.*, No. 11, 1991, pp. 48–51.

[19] Leopald, R. J., "The Iridium Communication System," *Proc. Singapore ICCS/ISITA '92,* Singapore, 1992, pp. 451–455.

[20] Usui, T., et al., "Satellite Constellations for a Multiple LEO Satellites Networks," *IEICE Trans.*, Vol. J75-A, No. 8, 1992, pp. 1370–1378.

[21] Kaniyil, J., et al., "A Global Message Network Employing Low Earth-Orbiting Satellites," *IEEE J. Select. Areas Commun.*, Vol. 10, No. 2, 1992, pp. 418–427.

[22] Chakraborty, D., "Survivable Communication Concept via Multiple Low Earth-Orbiting Satellites," *IEEE Trans. Aeroso. Electron. Syst.*, Vol. 25, No. 6, 1989, pp. 879–889.

[23] Katayama, M., and A. Ogawa, "A Study of the Communication Systems With Small Satellites," *Proc. 4th Symp. Small Satellites,* Tokyo, Japan, 1992, pp. 27–34.

[24] Ganz, A., Y. Gong, and B. Li, "Performance Study of Low Earth-Orbit Satellite Systems," *IEEE Trans. Commun.*, Vol. 42, No. 2/3/4, 1994, pp. 1866–1871.

[25] Katayama, M., A. Ogawa, and N. Morinaga, "Satellite Communication Systems With Low Earth Orbits and the Effects of Doppler Shift," *IEICE Trans. Commun.,* Vol. J76B-II, No. 5, 1993, pp. 382–390.

[26] Katayama, M., and N. Morinaga, "A Study of the Communication Systems Using the Low-Altitude Nongeostationary Satellites," *Proc. IEEE Int. Conference on Systems Engineering,* Kobe, Japan, 1992, pp. 452–456.

[27] Hashimoto, Y., et al., "A Study on a LEO Satellite Communication System for Experimental," *Proc. 4th Symp. on Small Satellites,* Tokyo, 1992, pp. 41–43.

[28] Hu, L-R., and S. S. Rappaport, "Personal Communication Systems Using Multiple Hierarchical Cellular Overlays," *IEEE J. Select. Areas Commun.,* Vol. 13, No. 2, 1995, pp. 406–415.

[29] Viterbi, A. J., "A Perspective on the Evaluation of Multiple Access Satellite Communication," *IEEE J. Select. Areas Commun.,* Vol. 10, No. 6, 1992, pp. 980–983.

[30] Bulloch, C., "The European Mobile Satellite Arena—INMARSAT, ALCATEL, QUALCOMM, and ESA," *Via Satellite,* Sept. 1992.

[31] Jamalipour, A., et al., "Signal-to-Interference Ratio of CDMA in Low Earth-Orbital Satellite Communication Systems With Nonuniform Traffic Distribution," *Proc. IEEE GLOBECOM Conf.,* San Francisco, CA, 1994, pp. 1748–1752.

[32] Jamalipour, A., et al., "Performance of an Integrated Voice/Data System in Nonuniform Traffic Low Earth-Orbit Satellite Communication Systems," *IEEE J. Select. Areas Commun.,* Vol. 13, No. 2, 1995, pp. 465–473.

[33] Jamalipour, A., et al., "Throughput Analysis of Spread-Slotted Aloha in LEO Satellite Communication Systems With Nonuniform Traffic Distribution," *IEICE Trans. Commun.,* Vol. E78-B, No. 12, 1995, pp. 1657–1665.

[34] Jamalipour, A., et al., "A Modified Power Control Scheme for Remedying the Effects of Traffic Nonuniformity in LEO Satellite Communications Systems," *International J. Wireless Information Networks,* Vol. 3, No. 1, 1996, pp. 29–39.

[35] Jamalipour, A., et al., "Transmit Permission Control on Spread Aloha Packets in LEO Satellite Systems," *IEEE J. Select. Areas Commun.,* Vol. 14, No. 9, 1996, pp. 1748–1757.

[36] Jamalipour, A., et al., "Adaptive Transmit Permission Control on Spread-Slotted Aloha Packets Applicable in LEOS Systems," *IEICE Trans. Commun.,* Vol. E79-B, No. 3, 1996, pp. 257–265.

[37] Jamalipour, A., and A. Ogawa, "Traffic Characteristics of LEOS-Based Global Personal Communications Networks," *IEEE Commun. Mag.,* Vol. 35, No. 2, 1997, pp. 118–122.

Selected bibliography

Bate, R. R., D. D. Mueller, and J. E. White, *Fundamentals of Astrodynamics,* New York: Dover Publications, 1971.

Jansky, D. M., and M. C. Jeruchim, *Communication Satellites in the Geostationary Orbit,* Norwood, MA: Artech House, 1987.

Jensen, J., et al., *Design Guide to Orbital Flight,* New York: McGraw-Hill, 1962.

Miller, M. J., B. Vucetic, and L. Berry, eds., *Satellite Communications: Mobile and Fixed Services,* Boston: Kluwer Academic Publishers, 1993.

"Mobile Satellite Communications for Seamless PCS," special issue of *IEEE J. Select. Areas Commun.,* Vol. 13, No. 2, Feb. 1995.

Pelton, J. N., and W. W. Wu, "The Challenge of 21st Century Satellite Communications: INTELSAT Enters the Second Millennium," *IEEE J. Select. Areas Commun.,* Vol. 5, No. 4, May 1987, pp. 571–591.

Pratt, T., and C. W. Bostian, *Satellite Communications,* New York: Wiley, 1986.

2

Communications with LEO Satellites

THE DESCRIPTIONS in Chapter 1 of different mobile satellite communication systems, and the disadvantages of geostationary satellite systems, make it clear that for future global PCNs utilizing very small hand-held terminals LEO satellites have a special position over other systems. This chapter examines LEO satellite systems and their specifications, including some practical, systematic problems with the LEO satellite systems and possible solutions.

Section 2.1 discusses some preliminary issues in the design of global LEO satellite systems. We start with the calculations of the number of satellites and the number of orbits required for a global coverage. We compare our numerical results to the ones specified in the proposals of LEO satellite systems and show how some of those designs assign some redundancies in their systems. Because the LEO satellites are in a continuous relative motion with the Earth, it would be necessary to change the connection of a user to different satellites, a procedure referred to as hand-off. The issue of intersatellite links, proposed, for example, in the IRIDIUM system, follow the discussion of hand-off. Intersatellite links may reduce the costs for terrestrial public telephony networks and make a satellite system completely independent of the ground facilities. Moreover, those links can make the communications system much more reliable in cases of disasters on the Earth, such as earthquakes. Such linkage is not, however, essential; many systems do not include it in their proposals, in order to reduce the total cost of their systems. The issues of the partitioning of satellites' footprints by spot-beam antennas and the Doppler shift effects in LEO satellite systems also are discussed in Section 2.1. The arrangement of spot beams is a new issue in satellite systems, which is opposite to the advantage of the wide coverage area often mentioned in satellite literature. The Doppler shift effect is a disadvantage for LEO satellite systems; their fast movement makes network control difficult.

Section 2.2 explains two issues in the implementation of LEO satellite systems. The first issue is the selection of a multiple-access scheme appropriate for LEO satellite systems. We discuss a group of multiple-access schemes and their possibilities for employment in a LEO satellite system. Two new and interesting multiple-access schemes—code division multiple access and spread-slotted Aloha—are introduced, and the reasons for employing those schemes are discussed. The selection of an efficient multiple-access scheme is an important issue in any wireless communications systems faced with large numbers of users utilizing a common channel. The second issue is the problem of geographical traffic nonuniformity, which does not affect conventional geostationary satellite systems, just LEO satellite systems [1,2]. Geographical traffic nonuniformity is an important problem, and ignoring it during system design leads to incorrect and unreliable system expectations.

Section 2.3 introduces some concepts for LEO satellite system models suitable for analysis of the performance of such systems. In particular, we define the different areas considered in LEO satellite systems, including service area, coverage area, and interference area. A useful model for a LEO satellite system should contain sufficient information necessary for analyzing the performance of the system and should be as simple as possible for mathematical tractability. The model should be competent with the realistic model of the LEO satellite systems, which often is very complicated. This chapter introduces a simple LEO satellite system model and shows how that simple satellite system model can offer almost the same qualitative characteristics as the complicated real system model with simpler mathematics.

2.1 Preliminary issues in LEO satellite systems

The design of a satellite system is a complicated, difficult task. It takes several years to design different parts of a system. A wide variety of engineers from many fields are involved in the design process, and for a satellite system for communications purposes electrical engineers are included in only one of those fields. Even that one field, a wide variety of electrical engineers are involved in different aspects of system design, such as electronics engineers for designing the power supply and electronic equipment, antenna engineers for designing transponder and wave propagation processing, and system and network engineers for the design of protocols, access schemes, and so on. Because this book is from an electrical engineering viewpoint, many issues not related to the electrical aspects of satellite systems cannot be discussed here. Even many subjects related to communications, for example, antenna gain, are not discussed in details here. Readers are referred to other books about such issues.

In addition to the difficulties in the design of general satellite systems, in the case of LEO satellite systems, a number of new problems occur. This section briefly discusses some primary design issues for these systems.

2.1.1 Required number of LEO satellites and orbits

It is obvious that a satellite at a higher altitude can provide wider coverage than one at a lower altitude. Chapter 1 mentioned that a geostationary satellite can cover about one-third of the globe; therefore, three satellites are sufficient for a global coverage. A LEO satellite system, on the other hand, requires tens of satellites for global coverage. In this section, we calculate the expected number of satellites and orbits for global coverage.

To calculate these parameters, assume a single satellite, as illustrated in Figure 2.1. Any location on the Earth can be seen from a satellite by

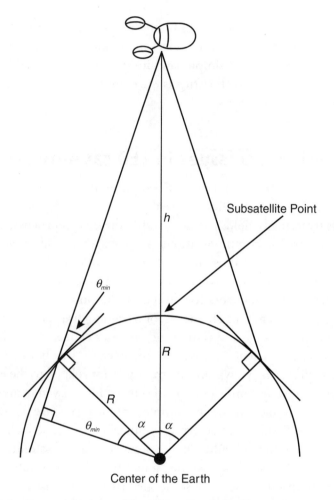

Figure 2.1 Geometrical considerations in a satellite system.

an angle, the elevation angle, denoted by θ. A minimum value of that angle, denoted by θ_{min}, defines the limits of the service area of a satellite. Because of the spherical shape of the Earth and natural obstacles, such as mountains, any location with an elevation angle less than θ_{min} cannot be easily seen from the satellite, so a reliable communication cannot be realized in very low elevation angles. For a single satellite, the circular footprint of the satellite as its service area, which means that any user located in the spherical segment of the Earth under the elevation angle of θ_{min} can be serviced by the satellite. The actual limits of the service area of a satellite are determined according to other parameters, such as the number of satellites and orbits in a satellite system.

According to simple mathematics, we can find the extent of the footprint of a satellite according to θ_{min} and its altitude, h. Then, by reference to the geometric relations shown in Figure 2.1, the half-sided angle of the footprint, α, measured at the center of the Earth can be given by

$$\alpha = \cos^{-1}\left(\frac{R}{R+h}\cos\theta_{min}\right) - \theta_{min} \tag{2.1}$$

where R is the average value of the radius of the Earth, about 6,378 km.

Now, we can use the value of α given in (2.1) to calculate a lower limit for the necessary number of satellites [3]. To cover the entire surface of the Earth, the footprints of the satellites should overlap. Without assuming any specific satellite constellation and to find the minimum number of satellites for global coverage, consider the largest possible effective footprint of a satellite as the largest hexagon inscribed into the footprint, as shown in Figure 2.2. Each hexagon consists of six isosceles spherical triangles, each with a central angle of 60 degrees and two identical angles ψ at the periphery of the footprints. Considering the spherical shape of the Earth (which is not well seen in the figure), the relation for the angle ψ is given by

$$\tan\psi = \frac{\frac{1}{2}\sqrt{3}\,\alpha}{\frac{1}{2}\alpha \cdot \cos\alpha} = \frac{\sqrt{3}}{\cos\alpha} \tag{2.2}$$

If σ denotes the spherical excess of the triangles, it is equal to

$$\sigma = 2\psi - 2\pi/3 \tag{2.3}$$

Then, the area of a hexagon will be given by

$$A_{hex} = 6R^2\sigma \tag{2.4}$$

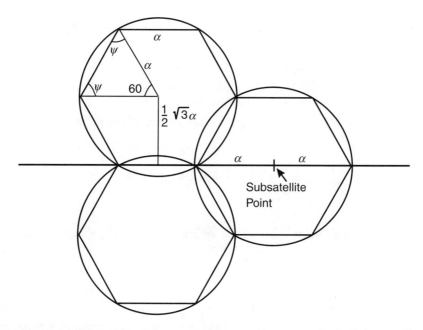

Figure 2.2 Geometric relations of the hexagons inscribed in footprints in global coverage.

Therefore, with the scenario shown in Figure 2.2, at least *n* satellites are necessary to cover the whole surface of the globe, where *n* is given by

$$n = \frac{4\pi R^2}{A_{hex}} = \frac{\pi}{3\psi - \pi} \tag{2.5}$$

Figure 2.3 plots the required minimum number of satellites for global coverage for different minimum elevation angles, according to (2.5). The figure also shows the number of satellites that are proposed by some LEO satellite systems and the GEO satellite system. The IRIDIUM system, as described in Chapter 1, has 66 satellites at an altitude of 780 km, with a minimum elevation angle of 8.2 degrees. GLOBALSTAR and ODYSSEY have 48 satellites[1] (at 1,400 km altitude) and 12 satellites (at 10,400 km altitude) with 20- and 30-degree minimum elevation angles, respectively.

---

1. Actually, GLOBALSTAR has 56 satellites; 48 satellites would be operational, and the other 8 would be on-orbit spares.

As can be seen in Figure 2.3, the number of satellites in GLOBALSTAR and ODYSSEY constellations is only slightly larger than the value we obtain from Figure 2.3. On the other hand, the number of satellites in the IRIDIUM system is much larger than the values we have obtained. One reason would be the assumption of minimum overlapping between the footprints of satellites in deriving (2.5). As mentioned in Chapter 1, in the IRIDIUM system, there is a lot of overlapping between service areas, to ensure more reliable communications.

The analysis presented here of the minimum number of satellites for global coverage can be sufficient, for example, in a GEO satellite system in which the satellites are in the plane of the equator and perfectly equispaced. In a real situation, orbits in general cannot maintain such a

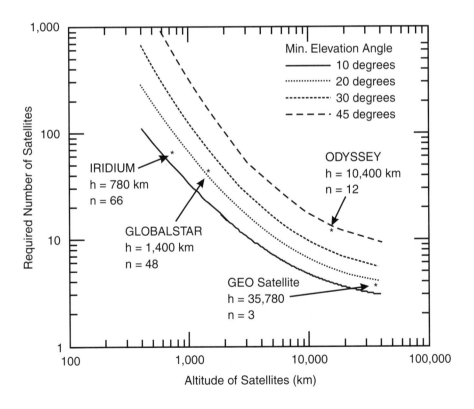

Figure 2.3 Required minimum number of satellites for a global coverage with different minimum elevation angles θ_{min}.

geometry, because the natural axis of symmetry is the rotation axis of the Earth. Thus, more complicated analysis is necessary.

For mobile satellite systems, besides the minimum number of satellites, the minimum number of orbits[2] of those satellites also must be determined. To determine the required number of orbits, we consider the coverage of a satellite at the equator. Such a situation is shown in Figure 2.2. Then, on the condition of at least two satellites in each orbit, each orbit would cover $3R\alpha$ of the equator. Hence, the minimum number of orbits for a global coverage is given by

$$\Omega = \left\lceil \frac{2\pi}{3\alpha} \right\rfloor \tag{2.6}$$

where $\lceil x \rfloor$ denotes the smallest integer equal to or greater than x. The minimum number of orbits for different orbit altitudes is shown in Figure 2.4. As can be seen in the figure, the resulting plots are very near to the values proposed by the LEO and MEO mobile satellite systems.

For a satellite system utilizing polar orbits, such as IRIDIUM, a more realistic estimation of the required number of satellites in each orbit can be given by

$$n' = \left\lceil \frac{2\pi}{\sqrt{3}\,\alpha} \right\rfloor \tag{2.7}$$

which results in a total number of satellites $\Omega \cdot n'$. For example, for an elevation angle of 8.2 degrees of the IRIDIUM, we have $\Omega = 6$ and $n' = 11$, which are the design parameters proposed by Motorola. It is noteworthy that some commercial proposals of big-LEO satellite systems consider larger numbers of satellites and orbits than the minimum requirements presented here, to make more than one satellite visible to each Earth user at any onetime. Although that increases the cost of the system, it also realizes some advantages, such as simplifying the hand-off process as a user moves from the service area of one satellite to the service area of

---

2. The minimum number of orbits are typically called planes because all orbits that follow that Keplerian path will be in a plane.

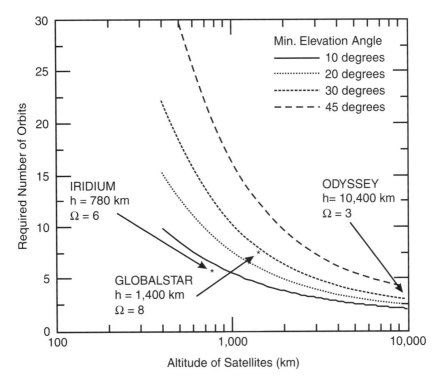

Figure 2.4 Required minimum number of orbits for global coverage with different minimum elevation angles θ_{min}.

another one and decreasing the shadowing effect from the obstacles in the line of sight of a satellite and a user.

2.1.2 Hand-off

In terrestrial cellular mobile communication systems, a number of fixed base stations and a large number of mobile users are involved. The service area of a base station is limited within an area where the level of signal power transmitted/received from/by that base station is higher than an acceptable value. In the first generation of cellular systems with high power transmitters at the base stations and large users' terminals, the service area of each base station was a circle with a radius on the order of 2 km. In recent cellular systems, called microcellular or picocellular systems, the radius is reduced to several hundred meters.

When a mobile user during an active connection leaves the service area of one base station and enters the service area of another, some mechanism must switch the old connection to a new connection without any interruption in the connection. Those mechanisms, referred to as hand-off protocols,[3] have been studied widely for cellular systems in last few years. It is obvious that as the radii of service areas become smaller, the number of hand-offs increases. For example, in the Japanese picocellular system, PHS (which has a large number of small base stations, each with a service area with a radius in the range of 100-500m), there would be many hand-offs during a typical 3-minute call. Of course the probability of changing the connection with base stations during a call is closely related to the speed of the mobile user. In a cellular system with high-speed mobile users, the probability is very high. In a system in which the users are mostly pedestrians, as in the case of PHS, the probability is low enough. In fixed telephony networks, the usual measure of the performance is blocking probability, that is, the probability with which a user wanting to make a connection to a desired destination, is refused because of loss of an idle channel in the system. In a mobile telephony system, in addition to that measure, we have a new measure related to hand-off. Sometimes this new measure is referred to as the "average number of hand-offs in a typical 3-minutes call" or as the "hand-off rate." An ideal hand-off protocol should be transparent to the users, who should not sense the change of connection to a new base station. If the number of hand-offs increased, the probability of interrupting the connection could be increased to unacceptable values because of nonideal protocols.

In the case of LEO satellite systems, the base stations (i.e., the satellites) move, and all mobile or fixed users on the ground appear almost as fixed users, because of the high ground speed of the satellites. In that case, a scenario similar to that of cellular systems exists; the only difference is that the mobility of users in a cellular system is replaced by the mobility of base stations in a LEO satellite system. For example, a LEO satellite at the altitude of 800 km in a complete orbit of the Earth requires about 100 minutes. In such a situation the largest visibility period of a

---

3. In the literature, the mechanisms also are called hand-over protocols. The term hand-off is used in analog systems, while hand-over is used in digital systems. This text uses the term hand-off, even though we sometimes are referring to a digital system.

given user to the satellite (depending on the minimum elevation angle) would be about nine minutes or less. Therefore, there would be the necessity of changing the connection between users and satellites during each connection with enough high probabilities. If the footprint of a satellite is partitioned into smaller cells, then, in addition to the hand-off between the satellites, hand-off between the cells of a satellite's footprint would be required as well.

A simple hand-off mechanism that is usually used in cellular systems and that would be applicable in LEO satellite systems, is having two upper and lower threshold bounds on the power of the received signals from users. Base stations continuously monitor the power of the received signals from users on uplinks. When the level of power of a user connected to a given base station falls below the upper limit, the base station assumes the user is moving to the next service area. At the same time, the base station of the new service area also senses the received power from that user above the lower bound. With the control channels between base stations, information on a connection can be exchanged between two base stations, and a new connection provided. In the case of a LEO satellite system with a large enough number of satellites, some areas commonly would be covered by two or more satellites at the same time; hand-off then can take place when a user is located in such areas. That way, the connections between satellites, which may be provided through the ground gateways or intersatellite links, have a dominant role in providing a reliable and successful hand-off without any sensitive call interrupt.

2.1.3   Intersatellite links

Another issue that has to be considered when a large number of LEO satellites are used in a system is establishment of a network between those satellites in order to interchange different kinds of information as well as route calls through a network in space. In the case of GEO satellite systems, in which the footprint of a satellite may cover several gateways on the Earth, such a network between satellites, or the *intersatellite link* (ISL), has not been considered so strictly. In GEO satellite systems, an intercontinental long distance call can be routed through gateways on the Earth; an additional cost for communications between satellites would not be necessary. However, for a LEO satellite at low altitudes and small coverage, all Earth gateways usually are out of sight of the satellite. An example of this situation would be a satellite over the Pacific Ocean.

Establishing a network between satellites is a complex and expensive task, because of large distances between satellites and the change of the relative positions of satellites, which requires antenna steering. Also, to establish ISLs, each satellite is required to have additional transmitters, receivers, and antennas, which increase the payload weight and the cost of the satellite. However, satellites in a system utilizing ISLs do not have to see the Earth gateway stations at all times, making the satellite system completely independent of the terrestrial facilities. Of course, because of economic considerations, all LEO satellite systems, including the ones that utilize ISL (such as IRIDIUM), propose internetworking with existing terrestrial systems. But the possibility of independence of the terrestrial systems is a great advantage of LEO satellite systems.

It often is said that since satellites are in the sky they can provide communications to the Earth even in the case of disasters. Experience, however, shows that that belief is not realistic. For example, at the time of a strong earthquake in Kobe, Japan, on January 17, 1995, four of the five GEO satellite gateway stations in the area were completely damaged, and the fifth one did not work because of the cut power supply, resulting in no communications for several hours. Providing voice communications in the area by portable satellite terminals took several hours and a lot of expense. In such situations, communication with satellites without any large and fixed gateways on the Earth is a great success, which can be provided by LEO satellite systems and their hand-held terminals, which utilize intersatellite links.

ISLs have been considered not only for LEO satellite systems but also for other systems in which different orbits are engaged. Some proposals for global communications propose a combination of satellites on low, medium, and even geostationary orbits. In such systems, for example, LEO satellites can cover local and dense traffics, and GEO (MEO) satellites can act as gateways or control stations for the LEO satellites and also cover wider areas and sparser traffics. In such systems, links between LEO and GEO (MEO) that can be termed as intersatellite links, allow for the exchange of information and control data between satellites of different orbits.

For a LEO satellite system, there are two types of intersatellite links, because the satellites are in several orbits. The first type is the link between satellites on the same orbital plane, called intraorbit intersatellite links. In this type, the satellites fly in a fixed successive arrangement, and the antennas can be almost fixed. The second type is a link between satellites on different (adjacent) orbital planes, called interorbit intersatellite links. Unlike intraorbit intersatellite links, in the second type, the ISLs require antenna steering. Moreover, because in the second type, the distances between satellites in different orbits vary within a large range and the Earth may interrupt their mutual line of sight, intersatellite communications have to be switched on and off at certain intervals. Figure 2.5 is an example of a LEO satellite system with the two types of ISLs.

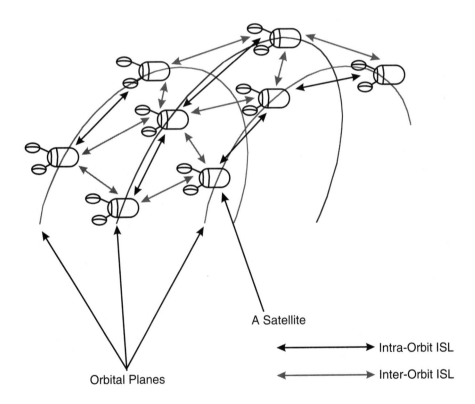

Figure 2.5 Two types of intersatellite links in a LEO satellite system.

IRIDIUM [4], for example, is a LEO satellite system that has proposed both types of ISLs (see Figure 1.8). Having an intersatellite network is a complex task that is very expensive, but it can provide reliable, Earth-independent personal communications in any part of the globe. A call started in a hand-held terminal on any part of the globe can travel long distances to another hand-held terminal through the networks of satellites and without any terrestrial facility. This may be one of ideals for the future PCNs that can be realized by LEO satellite systems [5,6].

2.1.4   Spot beams

One of the great advantages of a satellite system in communications is the wide area that can be serviced by a single satellite. In the case of conventional GEO satellite systems, that feature can provide long-distance intercontinental calls as well as real-time broadcasting throughout the entire globe. Even in the case of LEO satellite systems, in which the footprint of a single satellite is much smaller than that of a GEO satellite, it is still several times larger than the coverage area of a terrestrial station. A satellite can provide communications between two widely separated locations on the Earth, even when they are not in each other's line of sight due to spherical shape of the globe or geographical obstacles.

Although the wide-coverage feature is attractive, there are some disadvantages to it. For example, satellite signals ignore political boundaries as well as geographical ones, which may or may not be a desirable feature. Some countries do not like to receive special-program broadcasting via satellite signals (at least for public receptions). Also, transmitting signals to all parts of the footprint of a satellite requires a high level of power, provided by a satellite's batteries, power that is wasted in sparsely populated areas. In addition, many countries have their own satellites for national purposes, which do not require whole coverage in the footprint of satellites.

The concept of partitioning the footprint of satellites to more specific areas has been used for a long time in GEO satellite systems. In the case of GEO satellite systems, it is necessary to cover only the desirable areas for the purpose of a given satellite system and that enough number of gateway stations to be included in those areas. This can be done by arranging the projection of antennas on the satellite transponders to the desired areas.

In the case of LEO satellite systems, besides those requirements, which basically are designed for personal communications with hand-held terminals in addition to the communications of satellites with large gateway stations, it is preferable to have very small areas within the footprint of each satellite. Imagine the LEO satellite systems as an extension of the existing terrestrial systems, in which cellular-type base stations are replaced by LEO satellites [7]. In the case of terrestrial cellular systems, there are many activities to reduce the size of the cell into macrocells, microcells, and picocells, which can be applied in LEO satellite systems as well. In that manner, many big-LEO satellite system proposals consider the partitioning of the footprint of satellites into small areas, called cells, by very sharp spot-beam antennas on spacecraft. This is the concept of multicell LEO satellite systems. A simple illustration of such a configuration of small areas inside the footprint of a LEO satellite is shown in Figure 2.6.

By using multiple spot-beam antennas, the footprint of each satellite is divided into smaller areas, called cells (the name is borrowed from terrestrial systems). The basic concept of partitioning the footprint of a satellite into small cells came from the same idea used in current terrestrial cellular systems. Generally, as an advantage, with multiple-cell configuration, we can reuse the frequency bands in sufficiently separated cells or improve the bandwidth efficiency [8]. Another advantage is that by dividing a wide service area into small cells, the *radio frequency* (RF) power of users can be reduced due to the concentration of power into smaller areas, equivalent to a higher satellite antenna gain. With the power concentrated in small cells, the requirements for the variance in power of the user transmitters (which should be changed according to the location of user) can be more limited. With multiple spot-beam antennas, the power is reduced approximately by the number of cells in each satellite footprint. The issue of having lower transmission power is a key issue for realizing a global PCN with hand-held terminals.

As mentioned in Chapter 1 and to be discussed in more detail in Section 2.2, some big-LEO satellite systems, like IRIDIUM, proposed

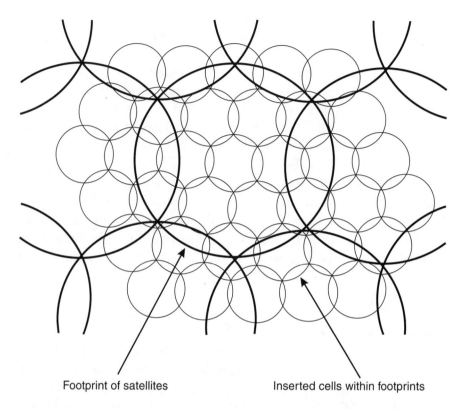

Footprint of satellites   Inserted cells within footprints

Figure 2.6 Partitioning of the footprint of a LEO satellite into smaller areas.

TDMA as the multiple access in their systems. Many others, like GLOBALSTAR, use CDMA, because CDMA exhibits attractive features in cellular systems that may be true for LEO satellite systems as well. The selection of a multiple-access scheme is closely related to the advantages of partitioning the satellite's footprint into small cells. We start the discussion of that selection with a review of the frequency considerations in the two systems.

In the IRIDIUM system proposal, the available 1616.0–1626.5 MHz band is divided into 31.5 kHz subbands with enough guard bands to space the individual carriers 41.67 kHz apart in a *frequency division multiplexing* (FDM) structure. In each 31.5-kHz subband, 89.96-ms TDMA frames are transmitted using *quadrature phase shift keying* (QPSK) modulation at a rate of 50 Kbps to form four uplink and four downlink channels. The

TDMA frame structure of the IRIDIUM is shown in Figure 2.7. In the figure, downlink (satellite-to-Earth direction) and uplink (Earth-to-satellite direction) channels are denoted by DL and UL, respectively.

In the case of GLOBALSTAR, a CDMA-based LEO satellite system, the basic structure is based on the one used in terrestrial CDMA systems of the *Interim Standard* (IS) 95. The frame structures on the uplinks and the downlinks of the GLOBALSTAR system between its mobile users and its satellites are shown in Figure 2.8. The frame consists of 1.25-MHz

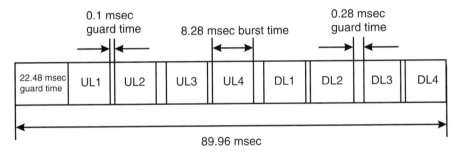

Figure 2.7. TDMA frame structure proposed for the IRIDIUM system.

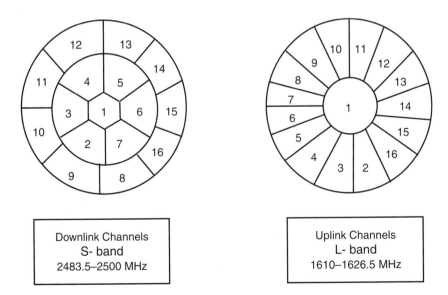

Figure 2.8 Downlink and uplink frame structures of the GLOBALSTAR system between mobile users and the satellites.

FDM subbands with FDD, so that the 1.25-MHz subbands required per traffic channel are paired. In this system, the 1.6-GHz band is subdivided into segments and used for transmitting each 1.25-MHz uplink. Similarly, the 2.4-GHz band is divided into an equal number of segments and is used for the corresponding 1.25-MHz band in the downlinks. As can be seen in Figure 2.8, the available frequency bandwidth in each direction is 16.5 MHz. In this system, there also are links between the satellites and its Earth gateways in C-band at the frequencies of 5091–5250 MHz and 6875–7055 MHz for uplinks and downlinks, respectively.

As mentioned, one advantage of partitioning the footprint of a satellite into smaller cells is the availability of frequency reuse in sufficiently separated cells, to improve the bandwidth efficiency of the system. In the case of TDMA systems, that can limit the level of interference from adjacent cells. Typically, a seven-frequency reuse pattern is used, whereby the central spot is surrounded by six spots whose FDM frequencies are different for each spot in any cluster of seven spots, as illustrated in Figure 2.9.

For CDMA, it is possible to reuse the same frequency in spots. That is because of the feature of CDMA in which the wideband signal mitigates multiple-access interference by using a correlation receiver that discriminates between desired and undesired signals. Therefore, it often is said that for CDMA the frequency reuse pattern is one, and then, the advan-

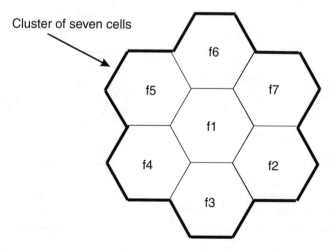

Figure 2.9   Seven-frequency reuse pattern in a TDMA system.

tage of spot beams in LEO satellite systems utilizing CDMA is not as strong as in the systems employing TDMA or FDMA schemes. As evidence of this, we can look at the parameters in proposals for the IRIDIUM and the GLOBALSTAR systems. The former one uses satellites at an altitude of 780 km, and the latter one uses satellites at 1,400 km. That means the footprint of a GLOBALSTAR satellite is much larger than that of the IRIDIUM. However, the GLOBALSTAR considers only 16 spot beams (fixed array) in each satellite footprint, compared to 48 for IRIDIUM. One reason for that is the less expected advantage due to spot beams in the case of CDMA and other spread-spectrum-based schemes. Section 2.2 discusses the advantages of spot beams in the case of CDMA-based systems, considering the traffic issues in LEO satellite systems.

When the footprint of a LEO satellite is divided into smaller cells, a kind of hand-off other than the one explained in Subsection 2.1.2 is necessary. In that case, when a user leaves one cell and enters another one, a new communication should be arranged, and the past connection canceled. Therefore, in a multicell LEO satellite system we have to arrange a hand-off mechanism between the cells within each satellite's footprint in addition to the hand-off required between the footprint of satellites, making the hand-off procedure more complex.

2.1.5 Doppler shift effect

Doppler shift is the problem of receiving higher or lower frequencies than the original transmitted frequency. It is caused by the relative high speed difference between a transmitter and a receiver. For example, if a transmitter with the frequency f moves very rapidly, a fixed receiver behind the transmitter receives the signal at a lower frequency equal to $f - \Delta f_1$, whereas another fixed receiver in the front of the transmitter receives the signal at a higher frequency equal to $f + \mathrm{D}f_2$. The change in the frequency, Δf, depends on several parameters, such as the distance between the transmitter and the receiver, the speed of the electromagnetic waves, and their relative velocity.

As shown in Chapter 1, in the case of satellite systems with low Earth altitudes, each satellite has a relatively high ground speed; hence, the system would be affected by high values of Doppler frequency shift. Figure 2.10 illustrates the scenario during the movement of a LEO

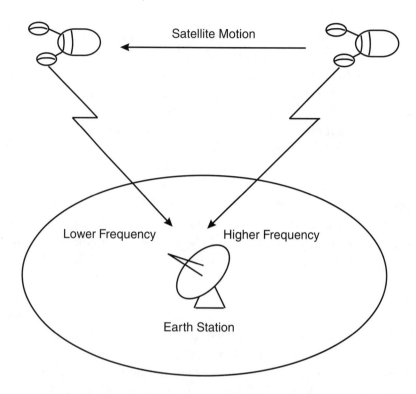

Figure 2.10 An illustration of the problem of Doppler frequency shift in LEO satellite systems.

satellite over a fixed ground station. As shown in the figure, as the satellite comes near the Earth station, higher frequency values will be reached at the station. On the other hand, as the satellite moves away from the station, signals at lower frequency levels will be received by the ground station.

Now assume the scenario illustrated in Figure 2.11. A fixed ground station is located on the equator, at a location denoted by "O," and a LEO satellite moves in a circle in the equatorial plane above that station and continuously transmits a signal at a given frequency, f_c. As the satellite moves over the station, different frequency values will be reached at the station. The change in the frequency of the signal, or the Doppler shift offset, relates to several parameters, such as the speed of the light, altitude of the satellite, rotational speed of the Earth, and the elevation angle at

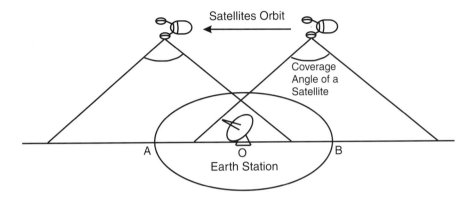

Figure 2.11 Schematic diagram of Doppler shift effects in LEO satellite systems.

which the station is seen by the satellite. The change in frequency can be given by a differential equation as [9]

$$\Delta f = \frac{1}{C} \frac{dD(t)}{dt} \tag{2.8}$$

where Δf is the Doppler frequency offset from the frequency f_c, C is the phase velocity of the light in free space ($3 \cdot 10^8$ m/s), and $D(t)$ is a time function for the distance of the satellite and the Earth station.

The shift in frequency due to the Doppler shift effect is closely related to the carrier frequency of the transmitted signal and the modulation method. Here, we assume a *binary phase shift keying* (BPSK) modulation, although similar results can be derived in the case of the more usual modulation technique in satellite systems, QPSK. In addition, to simplify the problem, we define the normalized Doppler shift as the value of Doppler shift offset divided by the carrier frequency. Figure 2.12 shows the normalized Doppler shift for different orbit altitudes, calculated with the carrier frequency and rate of 2.4 GHz and 19.4 Kbps, respectively. The horizontal axis in the figure is the time difference between the subsatellite point and the location of Earth station at the point denoted by "O." The time indicated as zero is the instant when the subsatellite point is at the location of the Earth station.

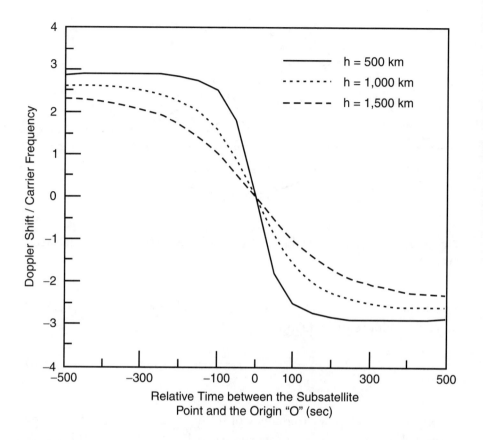

Figure 2.12 Numerical examples of Doppler frequency shift offset in LEO satellite systems at different satellite altitudes.

As can be seen in Figure 2.12, the change in the frequency shift becomes very sharp as the altitude of the satellite becomes low. The values of frequency shift in the case of circular orbit LEO satellite systems are much larger than the one that exists in a geostationary satellite system. On the other hand, in systems with highly elliptical nongeostationary orbits, satellites are used near their apogees, which are far from the Earth; hence, all the Earth stations in a service area have almost the same position relationships to a satellite and the same values of Doppler shift. In such a case, it is possible that the transmitters of the Earth stations obtain the value of the shift by receiving their own signals via the satellites and cancel the effects of Doppler shift by offsetting their carrier frequency with that value. However, in LEO satellite systems, because the distance between the Earth stations and the satellites is very small, some Earth stations suffer positive and others negative Doppler shifts at the same time. In that situation, the carrier frequency offset compensation by transmitters is impossible, and receiver-based techniques are required.

To solve the problem of Doppler shift for LEO satellite systems, several research studies have been done [9–11]. In those studies, it is shown that, for example, a receiver with *phase locked loop* (PLL) for wideband signal systems can be used to compensate the effects of Doppler shift. Also narrowband signals, another method in which the receiver uses the carrier regeneration by a block demodulator, can be more effective. The discussions given here and in the references imply that the effect of Doppler shift is a serious problem for the satellites on low Earth altitudes, but it can be solved by sophisticated design of transmitters and receivers.

2.2 Specific issues in LEO satellite systems

Some aspects of LEO satellite systems, such as calculations of the required number of satellites and orbits and the dynamics for the motion of satellites, are involved in preliminary stages of the system design. Some of them, such as the problem of Doppler shift, should be considered during the entire design, development, and maintenance of the system. In each area, many researchers and scientists from both the academic sphere and industry have been involved in the performance of LEO satellite systems, their constellation, their availability[4], and so on, since the early 1990s, when these systems were first commercially proposed. The quantity of those research studies exhibits the attractiveness of LEO satellite systems as strong candidates for future PCNs.

The main viewpoint of this book is the application of LEO satellites in global personal communication systems, so we will focus on two important issues in realizing such communications with LEO satellites. The first issue is the selection of a multiple-access scheme in a LEO satellite system that can distribute efficiently the limited frequency spectrum to the large number of users of a future personal communication

---

4. Availability in a communications system can be defined as the probability of attaining a specified level of performance and maintaining that level for a scheduled period of time.

system. The second issue is the problem of nonuniform distribution of the traffic loads around the world that should be serviced by the LEO satellite system. The second issue considered here is an original point of view concerning the topic of LEO satellite systems.

This section introduces those two issues. Although nomination of a specific scheme as the best multiple access in a system is a complicated task that requires consideration of many parameters and different traffic conditions, the author believes the schemes proposed here are appropriate candidates for such a nomination.

2.2.1 Selection of a multiple-access scheme

The selection of an efficient multiple-access scheme is an important issue in the design of any communications system that large numbers of users can independently access through a common channel with a limited frequency spectrum. For satellite communication systems of any types of orbit (GEO, MEO, or LEO), multiple-access techniques are essential; the problem is much more serious when the systems are employed for personal communications. That might be evidenced by the following facts: (a) the users (including users with either direct access to satellites by portable terminals or indirect access through gateways from public telephony networks) of mobile satellite systems are characterized by huge numbers; and (b) communications channels in mobile satellite systems suffer many imperfections, such as thermal noise, interference, fading, shadowing, and signal deflection. Accompanying those facts is the evolution of satellite access techniques from contentionless protocols to contention (random access) protocols and CDMA, which may be located between the other two protocols [12].

This section first reviews the issue of a multiple-access scheme and gives a classification for it. After that, it discusses multiple access for satellite communication systems.

2.2.1.1 Classification of multiple-access schemes

Before introducing the multiple-access scheme that will be used in this book, let us review the topic of *multiple access* and its main alternatives. Multiple access refers to the transmission of information data, which might or might not be inserted into packets, by numerous users to or through a common receiving point at the same time. A multiple-access protocol is a strategy to control such transmissions so that the probability of collision between them is maintained at a low enough level. The issue of multiple access with that definition can be applied not only to communication systems but also to computer networks in which computer terminals access a common processor of a mainframe or workstation. For that reason, different protocols have been invented.

As a general categorization of multiple-access protocols, we can divide them into contentionless and contention protocols [12]. In a contentionless protocol, users' transmissions are scheduled in either time or frequency domain. Therefore, sometimes contentionless protocols are referred to as scheduling protocols. By assigning a specific duration of time or frequency band to each user, it is possible to avoid collision between their signals. In that manner, fixed assignment schemes, such as *time division multiple access* (TDMA) and *frequency division multiple access* (FDMA), and demand assignment protocols have been proposed. In a demand assignment protocol, prior to the transmission, a user asks for a channel and after receiving the permission transmits a signal. In the fixed-assignment method, a user is allocated a part of channel capacity; in the demand-assignment method, the scheduling takes place only when the user has something to transmit, which can improve the spectrum efficiency of the system. In the latter method, no channel capacity is wasted on users who have nothing to transmit, which is unavoidable in the fixed-assignment methods.

In a TDMA scenario, each user is apportioned the entire transmission resource periodically for a fraction of time. Thus, for 10 users having equal requirements, 1 ms in every 10-ms frame might be assigned to each user. Each user's transmission, therefore, is intermittent, a condition that can be tolerated only by digital transmitters, which can store source bits and then burst them out at the transmission speed at which they were generated. Therefore, TDMA is employed mostly in digital transmissions. On the other hand, in an FDMA system, the allocated frequency spectrum is divided into subbands, and each user is apportioned a subband in the whole time domain. Different form TDMA, FDMA can be used with analog transmission as well as digital. Simple configurations of signals in TDMA and FDMA schemes are shown in Figure 2.13. Note that in either case it is necessary to insert some part of time or frequency

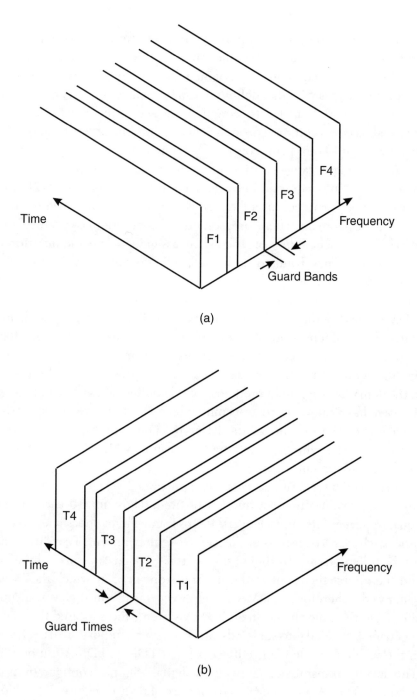

Figure 2.13 Frame structures in (a) FDMA and (b) TDMA schemes.

domain between adjacent channels in order to distinguish them from each other. The need for guard time in TDMA and guard band in FDMA schemes results in waste of some part of the precious resource.

In the contention protocol, users can transmit packets almost whenever they have something to transmit. This type of protocol sometimes is referred to as a random access protocol, because a user can randomly access the channel. Obviously, with random access, users cannot be sure that their transmissions will not collide with others; however, in this protocol, the synchronization control process is much simpler than the one required, for example, in a TDMA system. The most noteworthy of such protocols is the Aloha scheme, invented in 1970 [13], in which a user, without paying attention to the transmitting status of other users, can send packets completely randomly. Several protocols in the contention protocols group have been proposed since then, to improve significantly the low performance of the Aloha scheme. The Aloha scheme and its modified version, the slotted-Aloha scheme, are included in the repeated random access protocols subgroup. In the methods in that subgroup, it is necessary to retransmit all packets that have collided with each other. Another subgroup of random access protocols is those with reservation, in which a kind of reservation is necessary before the packets are transmitted. (Chapter 4 discusses Aloha schemes in more detail.)

CDMA schemes can be considered as either contentionless or contention protocols, depending on the situation of the channel [12]. A CDMA scheme is a contentionless protocol if the number of simultaneous transmissions on the channel or the level of multiple-access interference is under a given threshold in which all transmissions can be handled successfully. It is a contention protocol if the level of interference is above the threshold that results in contention and loss of all simultaneous packets.

In a CDMA system, users' signals occupy all of the frequency spectrum during the entire transmission period, but those signals are distinguished from one another according to the specific code assigned to each user, as shown in Figure 2.14. At any given time, a subset of the users in the system can transmit information simultaneously over the common channel to corresponding receivers. The transmitted signals in the common spectrum can be distinguished from one another by the superimposing of a different pseudo-random (or *pseudo-noise*, PN) pattern, called a

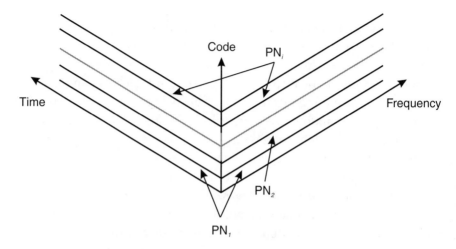

Figure 2.14 CDMA signals distinguished by PN codes.

code, in each transmitted signal. Thus, a particular receiver can recover the transmitted information intended for it by knowing the pseudo-random pattern, that is, the sequence used by the corresponding transmitter. The most popular form of CDMA is *direct sequence* CDMA (DS/ CDMA), in which DS spread-spectrum signals occupy the same channel bandwidth, provided that each signal has its own distinct PN sequence [14,15]. Because this form of CDMA spread-spectrum techniques and their performance enhancement through the processing gain are used for multiple-accessing purpose, it sometimes is referred to as *spread-spectrum multiple access* (SSMA).

The transmitter of each user in a DS/CDMA communication system typically has a structure like that shown in Figure 2.15. When several users transmit their packets to a common receiver with the structure shown in Figure 2.15, the difference in codes of the signals of each user makes it possible for the receiver to distinguish those signals from one another and to extract the individual information. In a spread-spectrum system, the narrowband signal of each user is multiplied by its PN sequence to form a wideband signal. If we assume that the information rate at the input to the encoder is R_b bps and that the modulation is BPSK, then the duration of a rectangular pulse corresponding to the transmission time of an information bit will be $T_b = 1 / R_b$, and the bandwidth expansion factor, or *processing gain*, of the spread-spectrum system will be

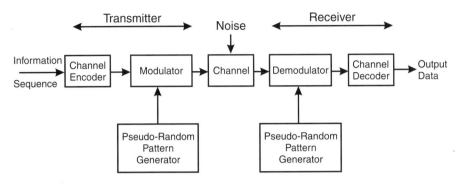

Figure 2.15 Model of the transmitter and the receiver for each user in a spread-spectrum multiple-access system.

$$B_c = \frac{W_{ss}}{R_b} = \frac{T_b}{T_c} \qquad (2.9)$$

where W_{ss} is the bandwidth of the spread signal and T_c is the reciprocal of W_{ss}, called the *chip interval*. A chip interval is defined as the time duration of the rectangular pulse (chip) of the spread-spectrum signal. The duration of a single bit in a DS spread-spectrum system then will be equal to B_c times the duration a chip.

2.2.1.2 Multiple-access schemes in satellite systems

Now we will discuss the issue of multiple-access schemes, especially CDMA, for mobile satellite communications, especially for LEO satellite systems. (The reader is strongly encouraged to see the literature [12,14–18] on spread spectrum and CDMA in order to understand the text that follows.)

The issue of multiple access first became important in communication satellites because of the need to utilize efficiently a precious spectrum resource. The first comparative study of three main multiple-access schemes—FDMA, TDMA, and CDMA—can be found in a paper published in 1966 [19] based on a digital communication satellite study performed for the U.S. Department of Defense.

Considering the limited applications of digital satellite communications through isolated-orbit geostationary satellites in 1966 and their large Earth terminals, that study proposed TDMA as the most appropriate scheme. The study mentioned that for the then-current applications of digital satellite communications, the most valuable commodity is the transmitted power of a satellite transponder since the satellite mass is proportional to payload power of the satellite. It is well known that to utilize the power of transmitters most efficiently, we must drive them into saturation area, where the amplifier operates as a nonlinear device. In that situation, if users' signals are being received simultaneously by the satellite on the uplinks, the nonlinear amplifier generates undesired intermodulation products, which both interfere with the desired signals and rob them of downlink transmitted power. In TDMA, because there is only one user accessing the satellite transponder at any given time interval, that problem can be avoided. However, a number of disadvantages offset that advantage. Perhaps the most important one is that the intermittent nature of the signal transmitted by the Earth stations requires a high peak-to-average power ratio, proportional to the number of users, which reduces the efficiency of the Earth transmitters.

An acceptable tradeoff between the advantages and the disadvantages would be large antennas and expensive high-power amplifiers on the ground. Therefore, the large trunking communication satellites launched since the 1970s and operated by the U.S. Department of Defense, INTELSAT and other satellites that interconnected very few large Earth stations, each having very high bit-rate requirements, mostly employed TDMA.

After the arrival of digital communication satellites networking large numbers of *very small aperture terminals* (VSATs) in the 1980s, however, the important consideration became the economics of the very many small Earth terminals, while satellite transponder costs could be amortized over a much larger terminal population. For those applications, TDMA was no longer employed, and most VSAT networks employed FDMA, with the more recent trends to employ CDMA [20–22].

Since the placing of satellites on lower orbit altitudes for commercial purposes, CDMA has become one of the strongest candidates for establishing multiple access in LEO satellite systems [23–25]. As mentioned in Chapter 1, many mobile satellite systems have been considered to employ CDMA schemes. While there does not appear to be a single multiple-accessing technique that is superior to others in all situations, there are characteristics of spread-spectrum waveforms that give CDMA certain distinct advantages. The two basic problems that face the mobile radio system designer are multipath fading of the radio link and interference from other systems. Spread-spectrum signals are effective in mitigating multipath because their wide bandwidth introduces frequency diversity. They also are useful in mitigating interference, again because of their widely spread bandwidth. The result of those effects is a higher capacity potential by employing voice activity and frequency reuse compared to that of nonspread access methods. Moreover, in CDMA, in contrast to FDMA and TDMA, integration of circuit-mode and packet-mode traffic requires no special protocol and makes an integrated voice/data system easy to realize; a multimedia feature of the future PCNs. In addition, in a packet radio environment, CDMA can offer uncoordinated random-access channel sharing with high throughput and low delay, along with other well-known advantages of spread spectrum, such as immunity to external interference and jamming and low probability of intercept.

Another alternative for the multiple-access scheme in LEO satellite systems is spread Aloha. This scheme is a combination of two well-known multiple-access schemes, CDMA and Aloha, which is also called CDMA Aloha [26] (see Chapter 4 for more details). Aloha by now is known as the simplest random access method that can realize the share of communications channels to a large number of users. In contrast to conventional (unspread) Aloha, in which the collision of the packets results in loss of the packets, if we spread the packets with pseudo-random sequences before transmissions, simultaneous multiple packet transmission only increases the level of multiple-access interference and does not always mean the loss of the packets: As long as the level of the interference is small enough compared to the strength of the power of the desired packet, the packet is accessible. In other words, in the spread Aloha scheme, we can receive a packet successfully while the level of the signal-to-interference ratio could be kept enough above a threshold value, thereby achieving a more improved throughput performance than unspread Aloha.

A subsequent scheme, spread-slotted Aloha [27], is a combination of CDMA and slotted Aloha that provides packet transmissions at specified time instants, allows us to have further efficient systems, similar to the relationship between the conventional (unspread) slotted and unslotted

Aloha schemes. We believe that for the LEO satellite systems, CDMA or its combined forms with Aloha, either slotted or unslotted, are appropriate schemes that can be flexible enough with the coming needs of future PCNs and their multimedia applications. Therefore, this book considers the performance evaluation of the LEO satellite systems employing CDMA and spread-slotted Aloha schemes. Chapter 3 presents calculation of the signal-to-interference ratio in a CDMA LEO satellite system; Chapters 4 through 7 employ spread-slotted Aloha and evaluate its performance. Those chapters also provide a number of techniques for improving the performance of the system.

2.2.2   Traffic considerations

Future global PCNs promise to be far more ambitious in terms of the number and the category of user terminals. Considering the current interests in personal communications services, it is expected that there will be an increase on the order of millions of subscribers, as mobile and personal services are provided by satellites. The service to subscribers provided by future satellite-based PCNs will not be limited to metropolitan areas and will be spread out around the world, including developing countries and even unpopulated areas.

It might be clear at this point that for future global personal communication systems, LEO satellites are the most promising candidates, because they can provide global coverage to small hand-held, low-power terminals without the need for any existing terrestrial facilities. LEO satellite systems, while having the most important features of conventional geostationary satellite communications systems, such as wide coverage area, direct radio path and flexibility of the network architecture, provide additional advantages for global communications networks, for example, low propagation delay, low propagation loss, and high elevation angle at high latitudes. LEO satellite systems, which can be considered an extension of cellular terrestrial systems, can provide all the necessities for tomorrow's PCNs. If we consider personal communications services to subscribers in unpopulated areas and in developing countries, on deserts and on oceans, what can be better than LEO satellites?

The LEO satellite systems proposed for future PCNs are designed to establish different kinds of services, such as voice, data, facsimile, and paging, to their consumer subscribers in all parts of the globe, including areas with relatively small numbers of users separately located in small cities. As a promising group of subscribers to these systems, we also can consider subscribers with dual-capability, hand-held PCN terminals capable of direct access to the satellite system as well as their source-country cellular system. The terminals of these users should have the capability of using the satellite system, because the people who carry such terminals can travel around the world, and the requirements of terrestrial systems in different countries are based on different standards. That dual capability will be one of the distinguishing features of future realistic PCNs.

As already mentioned, LEO satellite systems can be assumed to be an extension of existing terrestrial cellular systems, in which the cellular-type base stations are replaced by the satellites. Similar to cellular systems, in which several base stations are required to cover a specified area, for example, a country, tens of LEO satellites are necessary to realize global coverage. The reason is that the service area of a single LEO satellite is much smaller than the service area of a geostationary satellite, although it is wider than the conventional service areas of terrestrial cellular systems.

The service area of a LEO satellite may cover a number of small cities as well as urban areas; then the total traffic load of that satellite becomes much higher than that of its neighbor satellite, for example, one that is flying over an ocean. That traffic feature will be unique to LEO satellite-based PCNs. Generally speaking, that feature results in an important unbalanced traffic problem for the LEO satellite systems, one that requires intensive traffic engineering research. We refer to this issue as the problem of *geographic traffic nonuniformity*.

To manage global communications traffic, usually a model based on the extent of the coverage areas of different systems is offered. In this model, it is presumed that there is an internetworking between satellites in different altitude orbits and terrestrial systems. A sample of such management is shown in Figure 2.16. In this method, it is assumed that long-distance calls and communications between separately located areas are prepared via GEO satellites, whereas in more densely populated areas communications may be provided by LEO satellites and terrestrial systems. For personal communications in the next century, however, this model may not be any more useful, because users would communicate

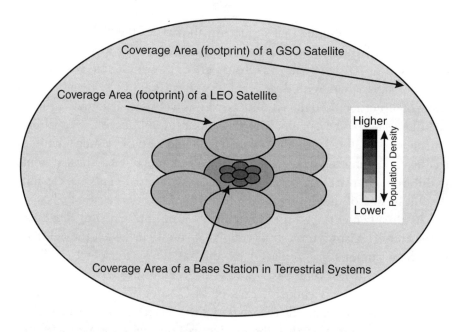

Figure 2.16 Assignment of different traffic densities to different communication systems.

mostly via LEO satellites and a cellular terrestrial system, where the latter is available.

It is a normal feature of a global PCN that different amounts of communications are requested in different areas. Only a few studies, however, have been done on the effect of nonuniform traffic distribution in LEO satellite systems. A few papers report some influences of traffic nonuniformity in terrestrial systems [28]. Although some concepts of these systems are applicable to LEO satellite systems [29,30], because of specific features of the LEO satellite systems, in the case of nonuniform traffic studies their results are not applicable to LEO satellite systems.

This problem basically is not the case for conventional geostationary satellite systems, since a geostationary satellite covers about one-third of the globe. However, for a LEO satellite system, in which the coverage of a single satellite can be as small as a part of a country or an ocean, the problem becomes apparent. The following chapters consider the geographical traffic nonuniformity problem in the LEO satellite systems and show how the performance of the system might be affected due to nonuniformity compared to a uniform traffic situation. We will consider a personal communication system in which the communications are provided only by LEO satellites, to exhibit the influence of geographic traffic nonuniformity on the performance of the system.

Before closing this introduction to the problem of traffic nonuniformity, it is interesting to consider the availability of the advantages of spot-beam antennas in nonuniform traffic distribution. The reason is that the advantages of utilizing spot-beam antennas in LEO satellite systems are very much effective in the traffic distribution of the system. Consider a satellite system in which the number of spot beams in a satellite footprint and available frequency bands are N_c and N_f, respectively. Then, in a narrowband system and uniform traffic assumption, the required system bandwidth will be reduced approximately by the factor N_c/N_f (an advantage for spot-beam architecture). However, in the case of nonuniform traffic distribution, we cannot expect the same advantageous effect. Especially if we assume the case in which all traffic channels are concentrated in a single spot beam, the gain in bandwidth efficiency is completely lost. On the other hand, in the case of spread-spectrum systems, after dividing the service area into multiple cells, the interference within the heavily loaded cell rises by a factor of N_c. Then, with respect to the fact that the spread-spectrum systems are interference limited, we should expect a link degradation of $10 \log N_c$.

This discussion shows that in the case of nonuniform traffic distribution, we cannot expect gain in bandwidth efficiency by partitioning the service area into cells, although the advantage of the reduction in the required transmitting powers of the satellites and users remains.

2.3 Modeling the LEO satellite systems

Now we will present a model for these systems that can be used in the mathematical evaluation of their performance. An appropriate model is one that contains, as much as possible, the necessary information for analyzing the performance of a LEO satellite system. The model should also be as simple as possible from the viewpoint of mathematical tractability. Moreover, the model should be competent with a realistic model of the LEO satellite systems, which often is complicated.

From the examples of LEO satellite systems given in Section 1.3, it is clear that for global coverage it is necessary to configure a number of LEO satellites on a multiorbit constellation, because of the small coverage of a LEO satellite compared to that of a conventional geostationary one. In such a multiorbit LEO satellite configuration, satellites exchange information and control packets with each other through gateway Earth stations or intersatellite links.

Therefore, in analyzing the performance of a LEO satellite system, we would be concerned with a relatively complicated multiorbit, multi-satellite global communications network. Assume that the satellites are on low, circular Earth orbits of the altitude h. The number of orbits and the number of satellites on each orbit are designed so that any area on the globe can be covered by at least one satellite at any given time. Users' terminals have the capability of direct access with satellites in both uplink and downlink directions. That assumption is realistic in the case of LEO satellite systems, because the low altitude of the satellites, that is, the low propagation distance between users and satellites, allows low-power, hand-held personal terminals. As a preliminary assumption, we also assume that a user communicates with the satellite that requires the lowest transmitting power, to minimize the total interference power on the channel. In a nonfading situation, that assumption means equal-size service areas for all satellites.

To establish a connection between a user and a satellite, it is necessary for the user to have an elevation angle larger than a minimum value of θ_{min} to that satellite. That minimum elevation angle provides an upper bound on the service area of each satellite, which will be determined for a satellite system according to the constellation parameters, such as number of orbits, number of satellites, and their altitudes, as well as the access method utilized in the system. We refer to the upper band of the service area to as the *coverage area* and to the actual one as the *service area*. With the assumption of equal-size service areas of satellites, the actual service area of each satellite has a hexagonal shape. If we consider the performance of a satellite system in which a type of spread-spectrum technique is employed for multiple accessing of users, the signal of any user located in the line of a satellite can reach that satellite and be added to the level of multiple-access interference at the satellite. Therefore, for any satellite, we should consider an *interference area*, the radius of which is defined by the elevation angle of zero.[5]

Figure 2.17 illustrates the different areas (coverage area, interference area, and service area) for a given satellite number *i*. In a LEO satellite system with a large enough number of satellites, some areas commonly

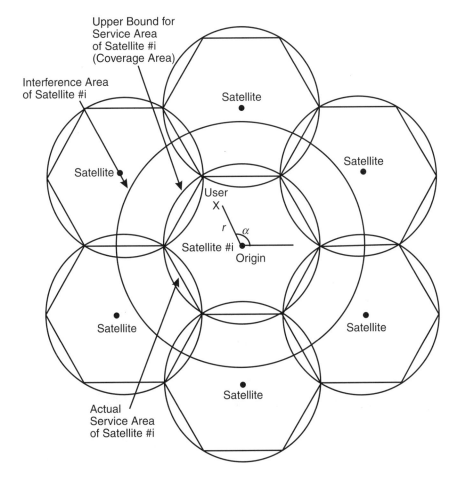

Figure 2.17 The areas in a LEO satellite system.

---

5. In some situations, even a zero-elevation angle may not be low enough. For example, the terrestrial microwave shows that there is ducting that will induce the waves to diffract to the heavier atmosphere below. That can allow propagation from beyond the line of sight.

would be located in the coverage area of two or more satellites. We refer to such areas, even those covered by more than two satellites, as *double coverage areas*. Figure 2.17 assumes that the globe is covered by a minimum number of satellites, so double coverage areas are limited to small areas. Table 2.1 summarizes those different definitions.

A realistic LEO multiple satellite communication system requires a three-dimensional model, but to make the mathematical calculations simpler and the effect of different parameters in the system clearer, we will consider one orbit of such a complicated system, represented by a two-dimensional model. Throughout the remaining chapters, it will become clear that the expectations of the two-dimensional model and the real three-dimensional one on the characteristics of the system have qualitatively almost the same tendencies. We will see that this model can exhibit well the degradation in the performance of the system due to the traffic nonuniformity as well as its performance improvement after the proposed schemes are applied.

In the two-dimensional model, each area on the Earth is represented by an arc measured at the center of the Earth, as shown in Figure 2.18. As illustrated in the figure, the different areas can be distinguished in the two-dimensional model as well. Although it is not shown in the figure, the actual service area of a satellite (the one in which users wishing to generate a communication select that satellite) may be smaller than its

Table 2.1
Summary of Different Area Definitions

Area	Description
Coverage area	An area with elevation angle to the satellite $\geq \theta_{min}$
Interference area	An area with elevation angle to the satellite ≥ 0
Service area	A defined limited area within a coverage area where users can connect to satellite
Double coverage area	An area commonly located between two or more adjacent coverage areas

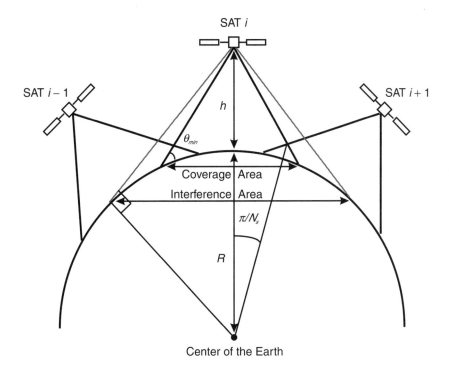

SAT *i*

SAT *i* − 1

SAT *i* + 1

h

θ_{min}

Coverage | Area

Interference | Area

π/N_s

R

Center of the Earth

Figure 2.18 A two-dimensional model of a LEO satellite system.

coverage area. One simple and natural example is the case in which all the satellites are equally spaced on a circular orbit and dedicated with the same-size service area. We refer to that method to as the natural method and its corresponding service area to as the *natural service area*, where with N_s satellites on each orbit, the size of the area measured by an angle at the center of the Earth is the same as the separation of the satellites, $2\pi/N_s$. It is worth mentioning that Earth users in the interference area but out of the coverage area of a given satellite would not be allowed to connect to that satellite, but their signals still can reach that satellite.

For most of the analyses in this book, we consider three succeeding satellites in the same circular orbit. The relations between the defined areas for those three satellites are shown in Figure 2.19. In the figure, the observed area is denoted as an area that is included in the interference areas of three succeeding satellites and will be considered in determining the performance of the LEO satellite system. We should note here that

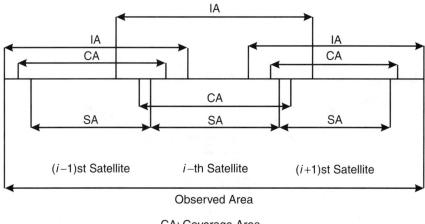

Figure 2.19 Illustration of the relations between service area, coverage area, and interference area.

throughout the analyses the performance of three succeeding satellites as a part of a global LEO satellite communication network is considered. That does not mean ignorance of the effects of other satellites in the same or adjacent orbits, for example, the ones shown in Figure 2.17. Because the interference area of each satellite is defined by the final line of sight of the satellite, the signal of any user in the interference area of that satellite can be reached at that satellite and adds to the level of its total interference at that satellite. We expect that partly examining the performance of the communication system can exhibit the performance of the whole system to some degree. In addition to the simplicity offered by the partial evaluation of the performance of the system, the effects of proposed schemes can be exhibited more clearly and their descriptions become more understandable.

2.4 Summary

This chapter introduced fundamental issues in the design of a LEO satellite communications as well as specific issues that are the focus of this book. The fundamental issues can be discussed in more detail from different viewpoints of engineering and not be limited to an electrical engineering perspective. The next chapter focuses the discussion on the performance of LEO satellite systems from a communications engineering viewpoint. The satellite system model introduced in Section 2.3 will be explained in more detail, and a traffic model that can exhibit the effect of traffic nonuniformity in the case of LEO satellite systems will be introduced. We will discuss the performance of the system when employing the two forms of spread spectrum, the CDMA and the spread-slotted Aloha, mentioned in Section 2.2, as promising candidates for the multiple access of users in future LEO satellite systems. As already mentioned, to follow the discussion in this book, the reader is expected to have some background in spread-spectrum systems, although some details are explained.

References

[1] Jamalipour, A., et al., "A Performance Analysis on the Effects of Traffic Nonuniformity in Low Earth-Orbit Satellite Communication Systems," *Proc. 16th Symp. Inform. Theory, Applicat. (SITA '93)*, Kanazawa, Japan, 1993, pp. 203–206.

[2] Jamalipour, A., et al., "Signal-to-Interference Ratio of CDMA in Low Earth-Orbital Satellite Communication Systems With Nonuniform Traffic Distribution," *Proc. IEEE GLOBECOM Conference*, San Francisco, CA, 1994, pp. 1748–1752.

[3] Werner, M., et al., "Analysis of System Parameters for LEO/ICO-Satellite Communication Networks," *IEEE J. Select. Areas Commun.*, Vol. 13, No. 2, 1995, pp. 371–381.

[4] Grubb, J. L., "The Traveler's Dream Come True," *IEEE Commun. Magazine*, No. 11, 1991, pp. 48–51.

[5] Chakraborty, D., "Survivable Communication Concept via Multiple Low Earth-Orbiting Satellites," *IEEE Trans. Aeroso. Electron. Syst.*, Vol. 25, No. 6, 1989, pp. 879–889.

[6] Kaniyil, J., et al., "A Global Message Network Employing Low Earth-Orbiting Satellites," *IEEE J. Select. Areas Commun.*, Vol. 10, No. 2, 1992, pp. 418–427.

[7] Jamalipour, A., and A. Ogawa, "Traffic Characteristics of LEOS-Based Global Personal Communications Networks," *IEEE Communication Mag.*, Vol. 35, No. 2, 1997, pp. 118–122.

[8] Pickholtz, R. L., "Communications by Means of Low Earth Orbiting Satellites," *Modern Radio Science 1996,* J. Hamelin, ed., Oxford Science Publications, U.R.S.I, pp. 133–149.

[9] Katayama, M., A. Ogawa, and N. Morinaga, "Satellite Communication Systems With Low-Earth Orbits and the Effects of Doppler Shift," *IEICE Transactions,* Vol. J76-B-II, No. 5, 1993, pp. 382–390.

[10] Katayama, M., A. Ogawa, and N. Morinaga, "Carrier Synchronization Under Doppler Shift of the Nongeostationary Satellite Communication Systems," *Proc. Singapore ICCS/ISITA '92,* Singapore, 1992, pp. 466–470.

[11] Katayama, M., and N. Morinaga, "A Study of the Communication Systems Using the Low-Altitude Nongeostationary Satellites," *Proc. IEEE Inter. Conf. System Engineering,* Japan, 1992, pp. 452–456.

[12] Prasad, R., *CDMA for Wireless Personal Communications,* Norwood, MA: Artech House, 1996.

[13] Abramson, N., "The ALOHA System—Another Alternative for Computer Communications," *Proc. 1970 Fall Joint Comp. Conf.,* 1970, pp. 281–285.

[14] Proakis, J. G., *Digital Communications, 2nd ed.,* New York: McGraw-Hill, 1989.

[15] Viterbi, A. J., *CDMA Principles of Spread Spectrum Communication,* Reading, MA: Addison-Wesley, 1995.

[16] Gilhousen, K. S., et al., "Increased Capacity Using CDMA for Mobile Satellite Communication," *IEEE J. Select. Areas Commun.,* Vol. 8, No. 4, 1990, pp. 503–514.

[17] Gilhousen, K. S., et al., "On the Capacity of a Cellular CDMA System," *IEEE Trans. Vehic. Technol.,* Vol. 40, No. 2, 1991, pp. 303–312.

[18] Kohno, R., R. Median, and L. B. Milstein, "Spread Spectrum Access Methods for Wireless Communications," *IEEE Commun. Mag.,* Vol. 33, No. 1, 1995, pp. 58–67.

[19] Schwartz, J. W., J. M. Aein, and J. Kaiser, "Modulation Techniques for Multiple-Access to a Hard Limiting Repeater," *Proc. IEEE,* Vol. 54, 1966, pp. 763–777.

[20] Viterbi, A. J., "A Perspective on the Evaluation of Multiple Access Satellite Communication," *IEEE J. Select. Areas Commun.,* Vol. 10, No. 6, 1992, pp. 980–983.

[21] Viterbi, A. J., "The Evaluation of Digital Wireless Technology From Space Exploration to Personal Communication Services," *IEEE Trans. Vehic. Technol.,* Vol. 43, No. 3, 1994, pp. 638–644.

[22] Abramson, N., "Fundamentals of Packet Multiple Access for Satellite Networks," *IEEE J. Select. Areas Commun.,* Vol. 10, No. 2, 1992, pp. 309–316.

[23] Kahn, R. E., et al., "Advances in Packet Radio Technology," *Proc. IEEE,* Vol. 66, 1978, pp. 1468–1496.

[24] Falk, G., et al., "Integration of Voice and Data in the Wideband Packet Satellite Networks," *IEEE J. Select. Areas Commun.,* Vol. SAC-1, No. 6, 1983, pp. 1076–1083.

[25] Wilson, N. D., et al., "Packet CDMA Versus Dynamic TDMA for Multiple Access in an Integrated Voice/Data PCN," *IEEE J. Select. Areas Commun.,* Vol. 11, No. 6, 1993, pp. 870–884.

[26] Sato, T., et al., "Throughput Analysis of DS/SSMA Unslotted Systems With Fixed Packet Length," *IEEE J. Select. Areas Commun.,* Vol. 14, No. 4, 1996, pp. 750–756.

[27] Makrakis, D., and K. M. Sundaru Murthy, "Spread Slotted ALOHA Techniques for Mobile and Personal Satellite Communication Systems," *IEEE J. Select. Areas Commun.,* Vol. 10, No. 6, 1992, pp. 985–1002.

[28] Sato, S., et al., "A Performance Analysis on Non-Uniform Traffic in Micro Cell Systems," *Proc. IEEE ICC'93,* Switzerland, 1993, Vol. 3, pp. 1960–1964.

[29] Pullman, M. A., K. M. Peterson, and Y. Jan, "Meeting the Challenge of Applying Cellular Concepts to LEO SATCOM Systems," *Proc. IEEE ICC '92,* Chicago, IL, 1992, Vol. 2, pp. 770–773.

[30] Chin, L., and J. Chang, "Using Low Earth Orbiting Satellites to Backup the Support of the Existing Ground Mobile Communications," *Proc. IEEE ICC '93,* Switzerland, 1993, Vol. 3, pp. 1103–1107.

Selected bibliography

"Code Division Multiple Access Networks I," special issue of *IEEE J. Select. Areas Commun.,* Vol. 12, No. 4, May 1994.

"Code Division Multiple Access Networks II," special issue of *IEEE J. Select. Areas Commun.,* Vol. 12, No. 5, June 1994.

"Digital Cellular Technologies," special issue of *IEEE Trans. Vehicular Tech.,* Vol. 40, No. 2, May 1991.

Glisic, S., and B. Vucetic, *Spread Spectrum CDMA Systems for Wireless Communications,* Norwood, MA: Artech House, 1997.

Miller, M. J., B. Vucetic, and L. Berry, eds., *Satellite Communications: Mobile and Fixed Services,* Boston: Kluwer Academic Publishers, 1993.

Pratt, T., and C. W. Bostian, *Satellite Communications,* New York: Wiley, 1986.

3

Application of CDMA in LEO Satellite Systems

A S WE HAVE SEEN, multiple-access schemes based on spread-spectrum techniques, especially CDMA, are promising candidates for future LEO satellite communications systems. As already mentioned, the reason that the wide bandwidth of spread-spectrum waveforms introduces frequency diversity, which can mitigate multipath and interference [1–6]. In addition, with CDMA it is possible to use the same carrier frequency in all service areas, so when a user leaves the service area of a satellite and enters the next area, switching the user's connection to a new satellite, a process referred to as hand-off, requires a simpler process than the parallel process in TDMA or FDMA. In this chapter, CDMA will be employed on the uplinks (i.e., users-to-satellites links) of our LEO satellite system, and performance of that scheme in uniform and nonuniform traffic distributions will be evaluated. In some literature, uplink is referred to as reverse link; throughout this text, however, we refer to it as uplink, because that name better indicates the physical direction of the link.

This chapter investigates the performance of a LEO satellite system for two types of traffic information scenarios. The first scenario is a general CDMA system designed to service analog-type terminals [7,8]. After that, we discuss the performance of CDMA in a LEO satellite system when an integrated voice/data traffic scenario is involved [9]. Examination of the first situation is made analytically, whereas the second system is described by simulation. Future LEO satellite-based PCNs are expected to support different types of information; hence, such an integrated information analysis is necessary.

This chapter will show that in analog systems when CDMA is applied on the uplinks, traffic nonuniformity causes large differences in the signal qualities at succeeding satellites; a satellite above a heavily loaded (dense) traffic area has a low *signal-to-interference ratio* (SIR), while its neighbor satellites over lightly loaded (sparse) traffic areas have a high level of SIR. As a result of that phenomenon, the performance of each user becomes a variable of location and the satellite with which the user is connected, not a proper feature in a communications system. To make the dependency weaker, a traffic assignment scheme, which makes the traffic load of the satellite over the dense traffic area smaller, is proposed and its performance improvement is estimated.

In the integrated traffic scenario, according to the simulation results, traffic nonuniformity affects the performance of the system almost in the same manner as for nonintegrated systems. It will be shown that the ratio of the population of data users to that of voice users has little effect on the performance of system. By modeling the satellite system during the movement of the satellites, the change in signal quality during peak traffic load in their route is also determined, and the worst case from the viewpoint of performance is derived.

The chapter is organized as follows. Section 3.1 considers an analog scenario. The mathematical model for the traffic nonuniformity and the calculations on SIR with a number of numerical examples are given in that section. Through the examples, we find the situation in which the traffic nonuniformity has the largest effect on the performance of the system and name it the worst case. At the end of the section, we propose a traffic assignment control method, which equalizes the traffic loads of service areas to some degree, and investigate its capability with numerical examples.

Section 3.2 continues the discussion of CDMA for an integrated voice/data scenario. We first explain the integrated voice/data system and the extension of the traffic model. The calculation of SIR and simulation environment also are given in this section. The performance of the system in both situations of the worst case and during the movement of the satellites is evaluated. The section finishes by proposing a modified power control scheme, very similar to the traffic assignment control scheme explained in Section 3.1.

3.1 Performance evaluation of analog systems

In this section, a general direct-sequence CDMA scheme is employed on the uplinks of the LEO satellite system, and its performance in uniform and nonuniform traffic distributions are discussed by the measure of SIR. Because in CDMA all users send their information with the same carrier frequency, the dominant factor that affects the signal quality is the interference from other users, rather than simple background (mostly thermal) noise, as in channel-assignment schemes such as TDMA or FDMA. Therefore, we use the expression "signal-to-interference" ratio instead of the conventional "signal-to-noise" ratio, although the background thermal noise is considered part of total interference.

We introduce the traffic model and then derive the SIR at individual satellites. In considering the movement of the satellites, we discuss the effect of relative locations of the satellites to the peak of the traffic load on SIR. We also investigate the performance variation according to the degree of the traffic nonuniformity.

3.1.1 Traffic modeling

Section 2.2 explained the existence of the traffic-nonuniformity problem in LEO satellite systems. According to the conclusions given there, we can expect the distribution of communications traffic loads on the globe to be a combination of heavily populated areas, lightly populated areas, and areas with very small population [10]. To model such geographic traffic nonuniformity mathematically, there might be different kinds of assumptions; for example, simply a rectangular pulse-shape traffic model in which the levels of the pulses show the levels of the traffic load at given parts of the globe. Another simple model may be a triangular-pulse model, in which the peaks of the triangles show densely populated areas. It seems that such a triangular model is much more realistic than the rectangular one, because sharp changes in the levels of the traffic load or in the number of users are not the case in the real world. The linear changes in the level of traffic that appear in the triangular model also seem not very realistic. If we accept having peaks of communication traffic load in some parts of the globe, much more realistic changes in the level of the traffic loads can be thought as normal or Gaussian shaped. These three possible shapes for a traffic model are shown in Figure 3.1. From the viewpoint of the total traffic load in a large-enough area, the triangular and normal models can offer the same results; however, from the viewpoint of the traffic loads in small areas, the normal model seems more familiar.

Another viewpoint in establishing a traffic model is its degree of simplicity during mathematical interpolations. Although a complex multipeak traffic model may show the real traffic-load distribution of the globe, the mathematics due to such model become complicated. A simpler single-peak traffic model can exhibit the most important effects of the traffic nonuniformity on the performance of a system. Moreover,

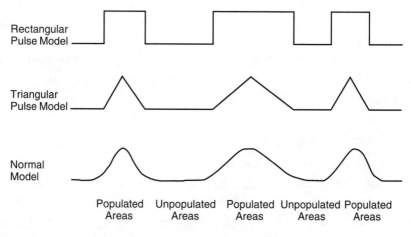

Figure 3.1   Examples of nonuniform-traffic models.

we can have a clearer discussion of the reason(s) for such effects and on the method(s) that can weaken those effects. With such an idea, we consider an area on the Earth equal in size to the summation of the interference areas of three successive satellites in the same orbit with a single peak of traffic load, as mentioned in Section 2.3.

Consider the circular LEO satellite system model explained in Section 2.3 and repeated here for convenience in Figure 3.2. To analyze the influence of the geographical nonuniformity of the traffic correspond- ing to the distribution of the users of the satellite system, total traffic load is modeled by a single-peak normal distribution of the population of the users in the observed area. The location of that single peak of the traffic load is assumed as the origin, and the location of any user is calculated according to this origin. The distribution of the users is assumed to have the following function [8,9]:

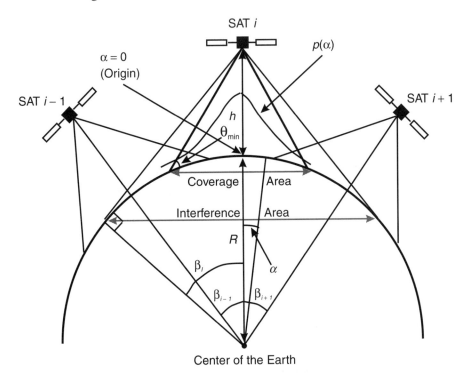

Figure 3.2 Typical shape of the normal nonuniform traffic model used in analysis.

$$p(\alpha) = \frac{A}{\omega} \exp(-\alpha^2/2\omega^2) \qquad |\alpha| \leq \pi \qquad (3.1)$$

where α is the angular distance of any user from the origin measured by the angle at the center of the Earth in radians, ω is a parameter representing the degree of uniformity of traffic, and A is a factor related to the total traffic load (total number of users) in the observed area (this factor will be explained later). With that definition, the traffic nonuniformity is expressed by ω; that is, larger values of ω expand the normal function more and an infinite value of ω realizes a uniform traffic distribution. A typical shape of $p(\alpha)$ is shown in Figure 3.2.

To investigate the effect of traffic nonuniformity with this model, the total traffic load for the satellites under consideration should be kept fixed when ω or the number of satellites in each orbit, N_s, changes. For that purpose, we assume that the total traffic load of three natural service areas, when the peak of the traffic is located at the origin, is constant and equal to B and thus

$$A = B \Big/ \int_{-3\pi/N_s}^{+3\pi/N_s} \Big[\exp(-\alpha^2/2\omega^2)/\omega \Big] d\alpha \qquad (3.2)$$

The ratio of the traffic loads of two adjacent natural service areas, when the peak of traffic is located at the origin, as shown in Figure 3.2, can be found from

$$\textit{traffic ratio} = \int_{-\pi/N_s}^{\pi/N_s} p(\alpha)d\alpha \Big/ \int_{\pi/N_s}^{3\pi/N_s} p(\alpha)d\alpha \qquad (3.3)$$

The traffic ratio for different numbers of satellites in an orbit, N_s, is shown in Figure 3.3.

3.1.2 SIR: The measure of performance

3.1.2.1 CDMA as multiple-access method and SIR calculations
This section evaluates the effects of traffic nonuniformity on the performance of the uplinks of a LEO satellite system. Direct-sequence CDMA is utilized as the multiple-access scheme in this direction.

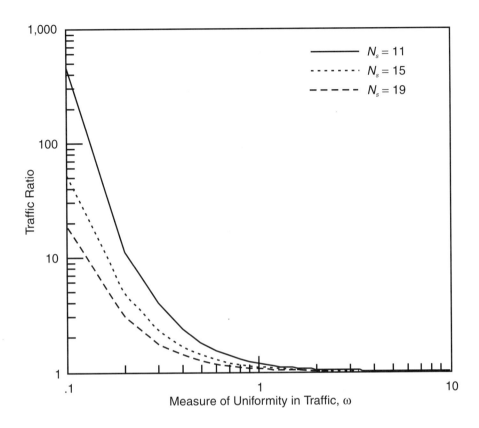

Figure 3.3 Ratio of traffic in the service area of two adjacent satellites (the first one is over the traffic peak) for different number of satellites in one orbit.

An important issue in multiple-accessing methods based on CDMA is the equalization of powers of receiving signals at each base station, referred to as reverse link *power control* [11–13]. With that control, the signals of all users reach the base stations with the same power regardless of their distances to the base stations. Without such control, the probability of successful transmission for users near the base stations is much higher than the ones far from the base stations. This phenomenon is referred to as the *near-far problem*, and it is said that power control is an effective method to remedy the problem. Moreover, it is said that power control can realize equal channel sharing and maximize the capacity of the CDMA.

In a LEO satellite system, this requirement can be satisfied if each Earth station has the knowledge of the required transmitting power levels to all visible satellites, by measuring the power of pilot signals received from the satellites [11]. Therefore, we assume that each satellite continuously transmits a pilot signal whose power level is known by all users. Comparing the received power level of this pilot signal with the referenced one, users can calculate their distance to all visible satellites and also can transmit their information with controlled power to allow reception at the connecting satellite with equal power to other users connecting to that satellite.

In the case of LEO satellite systems compared with terrestrial systems, because of relatively small effects of shadowing and Rayleigh fading, it is reasonable to assume that the radio signal power is attenuated in proportion to the second power of propagation distance [14,15]. (Shadowing and Rayleigh fading problems in the case of LEO satellite systems are discussed in Chapter 6.) Under that assumption, when the location of the Earth station (equipped with isotropic antenna) is α, the required transmitting power level to the ith satellite equals

$$P_i(\alpha) = \kappa S_i \cdot l_i^2(\alpha) \qquad i = 1, 2, \ldots, N_s \qquad (3.4)$$

where S_i is the designed receiving power level of the signals at the ith satellite, and $l_i(\alpha)$ is the distance (measured in meters) between the ith satellite and the Earth station at the angular offset α. κ is a constant with the dimension of m^2, but, as will be seen later, in final equations of SIR the ratio of powers rather than their absolute values are involved. Then κ can be assumed as unity, and hence we neglect it in the next equation for simplicity. (Note that if the background noise level is negligible compared to the level of multiple-access interference, κ can be completely neglected in the evaluation of SIR; however, if we consider the background noise as a part of total interference, κ cannot be neglected.) At the moment, we assume different power levels at satellites (e.g., S_i for the ith one) and derive a general form of SIR equations. However, by assuming $S_i = S$ for all i, an equal power scenario can be simply derived.

If the location of the ith satellite is β_i, the distance $l_i(\alpha)$ becomes

$$l_i(\alpha) = \left\{ \left[R + h - R \cos(\beta_i - \alpha) \right]^2 + R^2 \sin^2(\beta_i - \alpha) \right\}^{\frac{1}{2}}$$

$$i = 1, 2, \ldots, N_s \tag{3.5}$$

where R is the average radius of the Earth, about 6,378 km, and h is the altitude of the satellites. When two or more satellites are visible at the same time for an Earth station, the user compares the required transmitting power to each of them and connects to the satellite that requires lower power. Note that, at the instant shown in Figure 3.2, β_i is equal to zero, while the other two satellites have nonzero (positive and negative) values equal to the separation of the satellites.

When the uplink is designed to operate at an adequate power level, in CDMA systems, the effect of thermal noise generally is smaller than that of interference; hence, SIR is a proper measure of the system performance. Note that there is a simple relation between SIR and E_b/N_0; the ratio of signal energy per bit to interference plus noise energy (which sometimes is referred to as required SIR to achieve a specified error rate performance) is as follows [15]:

$$\text{SIR} = \left(E_b/N_0 \right) \left(\frac{2R_b}{W_{ss}} \right) \tag{3.6}$$

where R_b is the data bit rate, and W_{ss} is the spread-spectrum bandwidth. Then one can derive the error rate performance easily.

For a given user of the ith satellite, the transmitted signal arrives at that satellite with the power S_i, and signals of all other simultaneous transmissions from the users located in the service area and interference area of that satellite appear as additive interference. Thus, the SIR at the ith satellite becomes

$$\text{SIR}_i = S_i/l_i \tag{3.7}$$

where I_i is the total power of the interference at the ith satellite, described as

$$I_i = \int p(\alpha) \cdot \min\left[P_i(\alpha)\right] \cdot l_i^{-2}(\alpha)d\alpha, \quad i = 1, 2, \ldots, N_s \quad (3.8)$$

where min[x] is the minimum value of x and describes the connection of any user to the satellite that requires the lowest transmitting power. The interval of the integration in (3.8) should be determined for each area separately, as discussed next.

As a result of sphericity of the Earth, if the angular distance between a satellite and a user is larger than β_I, the transmitting signal of that user does not reach that satellite, where

$$\beta_I = \cos^{-1}(R/R + h) \quad (3.9)$$

is the interference limit angle. The interference limits are shown in Figure 3.2 for the ith satellite. Let all N_s satellites on an orbit request the same required transmitting powers, and thus users connect to the nearest satellite: natural service area configuration will be realized. In that case, from (3.8), the interference reached at the ith satellite at β_i can be determined from

$$I_i = S_i \int_{\beta_i - \pi/N_s}^{\beta_i + \pi/N_s} p(\alpha)d\alpha + S_{i-1} \int_{\beta_i - \beta_I}^{\beta_i - \pi/N_s} p(\alpha)l_{i-1}^2(\alpha)l_i^{-2}(\alpha)d\alpha$$

$$+ S_{i+1} \int_{\beta_i + \pi/N_s}^{\beta_i + \beta_I} p(\alpha)l_{i+1}^2(\alpha)l_i^{-2}(\alpha)d\alpha \quad (3.10)$$

where S_i, S_{i-1}, and S_{i+1} are the designed receiving signal powers at the ith and its adjacent satellites, which are now assumed to be the same values. The first term in (3.10) is the interference reached from the users located in the service area of the ith satellite; however, two other terms show the interferences from the users of neighboring satellites in the interference area of the ith satellite. If the designed receiving power differs for each satellite, the service area and the interference area will be different from those in (3.10). (The performance control that results from changing these powers is discussed in Section 3.1.3.)

3.1.2.2 The effect of satellite position on SIR

Because in a LEO satellite system the satellites are on nongeostationary orbits, they are in continuous motion, with relatively high ground speed, which is determined according to the altitude of the satellites and which was shown in Section 1.2 [16]. Here we examine the changes in SIR characteristics according to the travel of satellites, assuming that a non-uniform traffic distribution as (3.1), with a predefined value of ω in a specified area within the satellites' path, exists. In this analysis, we assume that the satellites are on circular orbits. The circular orbits usually are used in LEO satellite system constellations and simplify the control of the system considerably.

Figure 3.4 shows the changes in the SIR characteristics of two adjacent satellites, the ith and the $(i + 1)$st ones, as a function of β_i, that is, the angular position of the ith satellite, for typical constellation parameters of $h = 800$ km, $N_s = 11$ [17,18], and $\omega = 0.2$, when $S_i = S$ for all satellites. According to Figure 3.3, in this case, the number of users of the ith satellite is about twice those of its neighbor satellites, the $(i-1)$st and the $(i + 1)$st ones. Assuming counterclockwise rotation of the satellites, $\beta_i = -2\pi/N_s$ is the instant when the ith satellite is far from the peak of the traffic by equal angle as the separation of the satellites, $2\pi/Ns$, and the $(i + 1)$st one is just over this peak. They rotate in their circular orbit with a constant angular velocity, until the ith satellite reaches a symmetrical position to the traffic peak as the start point, that is, $\beta_i = 2\pi/N_s$. At the halfway point, when $\beta_i = 0$, the ith satellite is located just above the traffic peak. If we define the worst situation of system performance as the case when the signal quality at a satellite has the lowest value, from this calculation, this is the case where the peak of the traffic load lies just under one of the satellites, that is, $\beta_i = 0$ and $\beta_i = -2\pi/N_s$. Figure 3.4 shows that at those points the satellite above traffic peak has low signal quality; however, its neighbor satellite has a large value of SIR.

An important point drawn from Figure 3.4 is that there are large variations in signal quality at each satellite when a nonuniform distribution of users exists. That phenomenon may be acceptable while the level of SIR is higher than a threshold that ensures an acceptable error rate;

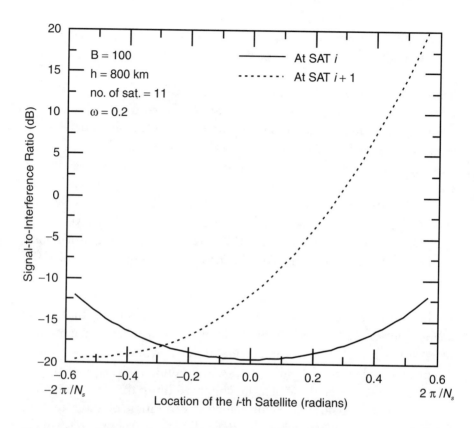

Figure 3.4 Changes in SIR characteristics of two neighboring satellites as a function of their angular locations.

however, it means that the users of the communications system have to accept large tolerances in their service quality performance, even during short periods of time, which is not good behavior for a reliable communications system.

3.1.2.3 SIR and Traffic Nonuniformity

According to the definition of the worst case given in the Subsection 3.1.2.2, here we examine how the degree of traffic nonuniformity affects the performance of the system. In the case where one satellite, say, the ith one, is above the traffic peak, SIR at the ith satellite and at its neighbor, the $(i+1)$st satellite, become

$$\text{SIR}_i = \frac{1}{2} \left[\int_0^{\pi/N_s} p(\alpha)d\alpha + \int_{\pi/N_s}^{\beta_I} p(\alpha) \cdot I_i^2 {}_{+1}(\alpha) \cdot \overline{I_i}^{-2}(\alpha)d\alpha \right] \quad (3.11)$$

$$\text{SIR}_{i+1} = 1 / \left[\int_{\pi/N_s}^{3\pi/N_s} p(\alpha)d\alpha + \int_{2\pi/N_s - \beta_I}^{\pi/N_s} p(\alpha)I_i^2(\alpha)\overline{I_i}^{-2}{}_{+1}(\alpha)d\alpha \right.$$

$$\left. + \int_{3\pi/N_s}^{2\pi/N_s + \beta_I} p(\alpha)I_i^2 {}_{+2}(\alpha)\overline{I_i}^{-2}{}_{+1}(\alpha)d\alpha \right] \quad (3.12)$$

again with the assumption that the satellites have the same designed receiving power levels, S. For the sake of simplicity, let us name the ith satellite above the peak of the traffic with a large number of users as the *dense traffic satellite* (DTS) and its neighbors with smaller numbers of users as *sparse traffic satellites* (STSs). Figure 3.5 shows the SIR characteristics at the DTS and each STS as a function of traffic nonuniformity for $h = 800$ km and $N_s = 11$. As can be seen in the figure, in large traffic nonuniformity (i.e., small ω), there are large differences between the signal qualities at the satellites, one above the dense traffic area and another above the area with sparse traffic. Also, the result of the case when more satellites exist in each orbit is shown in Figure 3.6. When the number of satellites in each orbit is increased, the service area and hence the number of users connecting to each satellite is decreased. It seems that the increase in the number of satellites can improve the performance of satellites with high communications traffic; although such methods increase the total cost of the satellite system. However, from Figure 3.6, we can conclude that the increase in the number of satellites gives negligible performance improvement for the DTS when the degree of traffic nonuniformity is large. For the case in which the number of satellites is 11 but the satellites have a higher altitude (i.e., $h = 1,500$ km), almost the same result as Figure 3.6 can be achieved; thus, the same kind of conclusion on the effect of the altitude on SIR can be drawn.

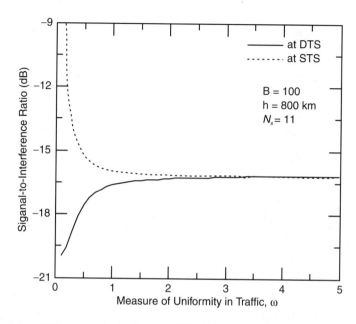

Figure 3.5. SIR characteristics at DTS and STS with the same required transmitting power levels, for $N_S = 11$.

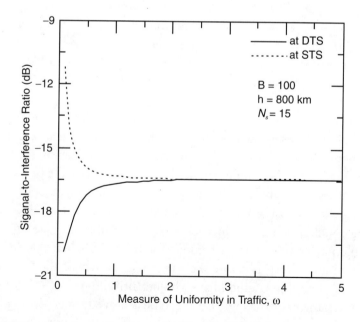

Figure 3.6. SIR characteristics at DTS and STS with the same required transmitting power levels, for $N_S = 15$.

As mentioned in Chapter 2, in the LEO satellite systems, some regions are in the coverage area of two or more satellites at the same time. Those areas, conventionally referred to as double coverage areas, are the result of the altitude and the number of satellites in each orbit. In a system with satellites higher in altitude or with a larger number of satellites in each orbit, the portion of the users located in double coverage areas becomes larger; hence, we observe some performance improvement. Note that a user in a double coverage area has the choice to connect to any the satellites that cover the area, if the protocol used in the system permits such selection to its users. The existence of such double coverage areas in LEO satellite systems suggests that there should be some flexibility in the definition of a service area other than natural method. It also suggests the possibility of performance improvement with other methods of assignment to users to satellites.

3.1.3 Traffic assignment control

3.1.3.1 Optimum control

Subsection 3.1.2 assumed that the required receiving powers of all satellites are the same and, hence, that service areas of all satellites are equal in size, referred to as natural service area configuration. That configuration, although natural in the case of uniform traffic, no longer has merit when the nonuniform distribution of users is involved. There needs to be a method that can change the size of service areas according to the offered traffic loads. As an example of such a method, this section proposes a scheme in which the designed receiving powers of the satellites are not equal. The proposed method would control the size of service areas according to their local traffic loads; that is, the service areas with light traffic loads are expanded, and the ones with heavier traffic loads are decreased. Obviously, in the case of uniform traffic, the size of service areas would be returned to the areas that appear in a natural service area configuration.

In this manner, let us first assume that the peak of the traffic is located under the ith satellite, that is, the DTS. Because the users communicate with the satellite that needs the smallest transmitting power, by increasing the required transmitting power of the DTS compared with its adjacent satellites on both sides (the STSs), it is possible to increase the tendency of the users in a double coverage area to connect to the STS, not to the DTS, thus decreasing the traffic load of DTS. This method is realized by changing the ratio of designed receiving power of the DTS to that of its neighbors on both sides, say, increasing the ratio $\gamma = S_i/S_{i-1} = S_i/S_{i+1}$ (which was unity in the last section). Each satellite counts the number of its users in a given period of time and by the means of intersatellite links, for example, the numbers of users of individual satellites are compared with each other, and then the proper ratio of γ in each area for the next period of time is selected and established. Figure 3.7 shows an example of the change in SIR as a function of the ratio γ in a relatively large traffic nonuniformity situation of $\omega = 0.2$. By increasing the ratio γ from 1, the number of users of the DTS and STSs is increased and decreased, respectively. Therefore, the performance of the DTS gradually improves and those of STSs degrade. As shown in Figure 3.7, as the ratio of the designed receiving powers, γ, increases, the SIR curves reach to a cross-point. Increasing the powers ratio more, makes the performances

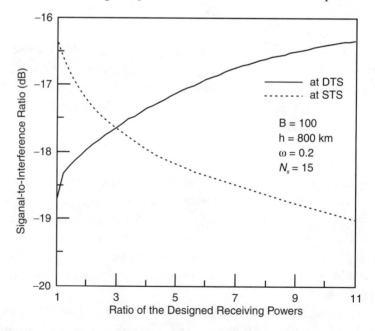

Figure 3.7 Changes in SIR characteristics at DTS and STS as a result of a change in the ratio of the designed receiving power levels at the satellites.

of STSs worse than that of the DTS. From the viewpoint of service quality for a given user, it is desirable to establish the same signal qualities when the connecting satellite of a given user is changed. Therefore, if we define the optimum control as one that makes all the satellites have the same SIR characteristics, the cross-point exhibits the optimum control.

3.1.3.2   Measuring the optimum capability
Taking appropriate powers ratios that achieve the optimum control for each traffic nonuniformity situation (i.e., for each ω) gives the SIR characteristics shown in Figure 3.8 and Figure 3.9 for $N_s = 11$ and $N_s = 19$, respectively. These figures show the changes in SIR characteristics at DTS and STS before and after applying the control scheme, where the solid lines show the SIR at both the DTS and the STS when optimum control

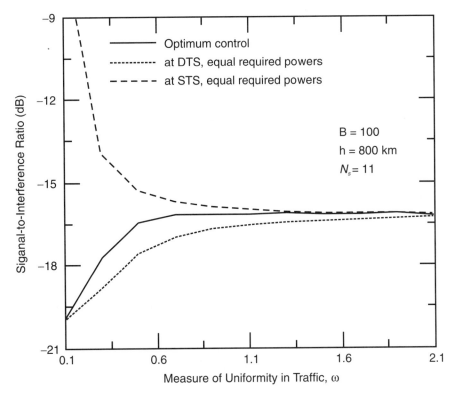

Figure 3.8   Effect of controlling the required transmitting powers of satellites in its optimum case, for $N_S = 11$.

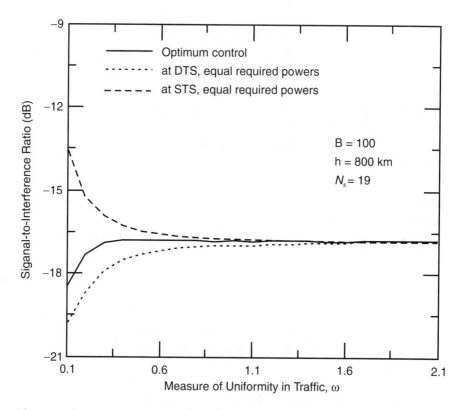

Figure 3.9 Effect of controlling the required transmitting powers of satellites in its optimum case, for $N_s = 19$.

is employed. As the figures illustrate, by applying this method, we are able to improve the performance of the DTS, that is, the satellite with heavy traffic load. There are two reasons for that improvement. First, by increasing the required transmitting power of the DTS, the service area of the DTS becomes smaller, and thus interference from its own users decreases. Second, lower transmitting powers of the users of STSs make the interference power from their users smaller. The latter occurs for every case, and if the double coverage area becomes large, the former effect also can be expected. Thus, as in the case in Figure 3.9, in which a larger number of satellites in each orbit is considered and then all areas on the ground are covered by at least two satellites, it is possible to improve the performance of the DTS even for a very small value of ω, such as 0.1.

Let us now change the relative position of satellites to the peak of the traffic load or the origin. Figure 3.10 shows another example, where the peak of the traffic is not just under one satellite: the ith and the $(i + 1)$st satellites are at $\beta_i = -\frac{1}{4}(2\pi/N_s)$ and $\beta_{i+1} = \frac{3}{4}(2\pi/N_s)$, respectively. In this case, even with the small double coverage area as occurred in the case of $N_s = 11$ and with small ω, the optimum control can improve the performance of the satellite that has the larger traffic load. It should be mentioned that any improvement in performance of a satellite requires a large degradation in performance of its neighbor satellites; however, with a trade-off between the achievable performance improvement and the related numbers of users of the satellites, the method becomes more attractive. That is, with this method, we improve the signal qualities for

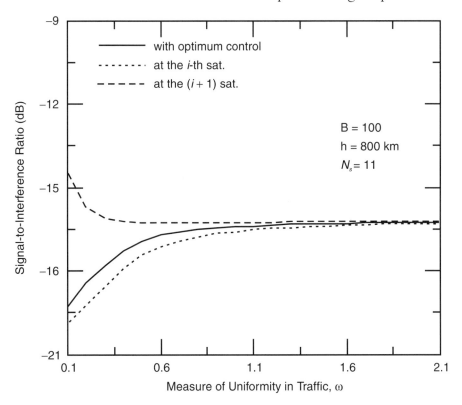

Figure 3.10 Effect of controlling the required transmitting powers of satellites in its optimum case, for $N_s = 11$ and different satellite positions.

a large population of users at the expense of degrading the performance of a smaller population of users. The important point is that we can maintain the performance for all users at an acceptable level; establishing the same signal quality at all service areas is much easier.

3.2 Performance of integrated voice/ data systems

In Section 3.1, the effect of traffic nonuniformity on the signal quality and performance of a LEO satellite communications system was determined. That analysis made no distinctions on the information type. In this section, we plan to examine the problem of traffic nonuniformity in an integrated voice/data scenario, which is of much interest in future PCNs [19,20].

To follow the calculations given in Section 3.1, in this section we determine the SIR characteristics at LEO satellites through simulation in two steps [9]. First, the case in which the satellites are assumed to be fixed with respect to the heavy traffic area in a short period of time is considered, and the relation between their performances and the intensity of traffic nonuniformity is estimated. After that, the investigation is generalized to the real case, that is, during the movement of satellites. Although the main purpose here is the estimation of the performance of LEO systems in nonuniform traffic situations, a modified power control method with the aim of remedying the effects of traffic nonuniformity also are discussed.

3.2.1 System considerations

Consider again the LEO satellite system model explained in Section 3.1.1. For such a system, we consider the effects of traffic nonuniformity on the performance of the system according to its uplinks. In this direction for multiple accessing by users to the satellite channel to transmit their packets, a packet CDMA scheme is used. We assume a simple CDMA protocol, in which all user information is transmitted in the form of a sequence of fixed-length packets on the channel. Access to the satellite channel is completely unconstrained (i.e., random access), so that any given users begin transmission whenever they are ready to send data. Moreover, assume there is no restriction on the transmitted-information type, which can be either voice or data. Data traffic is sent out as a single contiguous burst at the available peak CDMA channel speed, as in Aloha channels [5,21], packets not received successfully at the satellite are retransmitted repeatedly (after appropriate random delay) until an acknowledgment eventually is received. On the other hand, constantly generated bit-stream traffic such as voice is sent as a periodic sequence of packets with the duty cycle adjusted to match the requirements of the constant bit-rate source. Stream traffic normally cannot be retransmitted, so the receiver has to accept the packet loss rate caused by multiuser interference. More details about realizing this kind of mixed voice/data traffic scenario can be found in some papers (e.g., see [19,20,22,23]) and explained in the following.

In CDMA, in contrast to circuit-switching methods, integration of circuit-mode and packet-mode traffic requires no special protocol structure. On the other hand, in CDMA, users' transmitter powers should be controlled in such a manner that the received powers at the satellite become constant, avoiding the inbound channel receiver by close-in transmitters. As mentioned, in LEO satellite systems after the signal at the satellite is despreaded, all the simultaneous transmissions from the users located in the interference area of the satellite appear as additive interference. In this section, it is assumed that this kind of power control has been perfectly employed. It also is assumed that the uplink is designed to operate at an adequate power level, so that thermal noise effects need not be considered in the capacity and performance model.

Voice and data messages are formatted into packets as illustrated in Figure 3.11. As can be seen in the figure, each voice packet contains, in addition to its information bits, a synchronization preamble, network header, and bit error correcting code. The continuous bit stream of voice is broken up into periodically spaced packets, each with header, synchronization and error correction overheads. Here, it is assumed briefly that the channel transmission speed, R_t, exceeds the voice encoder bit rate, R_v; hence, the required duty cycle for transmission will be about R_v / R_t.

The actual packet size depends on which error correction method has been used. For example, with *Bose-Chaudhuri-Hocquenghem* (BCH) coding, which is capable of correcting n bits of error, the packet size, L, and the number of the bits of information, including the network header,

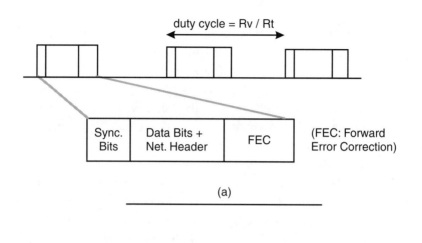

(a)

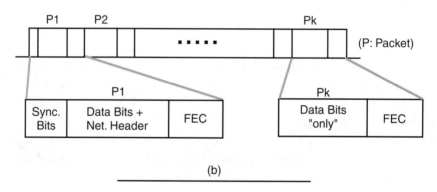

(b)

Figure 3.11   Formatting of (a) voice and (b) data packets for transmission on uplinks.

residing in data field, N, satisfies the relation of $N = L - n \log_2 L$ [15,22], where $n \log_2 L$ gives the number of bits of the error correction field of the transmitting packet.

In the case of data packet transmission, because the data message is transmitted as a contiguous sequence of L bit packets, the header and the synchronization bits are necessary to transmit only at the beginning of the messages. Therefore, in this case, only the data field of the first packet contains the network header bits, and, similar to voice packets, all the data fields of the packets have N bits of data. Unless the acknowledge signal for correct reception of entire packets of message is received by the user, a data message is assumed to be in error; hence, after an appropriate time-out, an Aloha-type retransmission procedure is executed by the transmitter. This procedure executes until the message is successfully received and acknowledged by the connected satellite.

Because the connection to LEO satellites for any ground user is temporary due to the relative movement of satellites, if during the transmission of packets or before receiving acknowledgment the user is forced to change connection to a new satellite, the information on the past connection is forwarded to the new satellite. After that, the new satellite should handle the process of transferring the user's packet from the old satellite or the transmission of an acknowledgment packet to the user. Such a mechanism should be prepared in the hand-off process. In the case of satellite systems with intersatellite links, the information can be easily exchanged via those links. In systems without intersatellite links, the information might be exchanged via ground gateways.

3.2.2 Extension of the traffic model

To apply the calculations given in Section 3.1 to an integrated voice/data scenario, we extend the traffic model of the system, keeping the nonuniform distribution of the users same as before. The population of the users is divided into two sets: voice users, $N_{v_i}(t)$, and interactive data users, $N_{d_i}(t)$, where the subscript i denotes their relation to the ith satellite, measured at the time of process, t. Each user lies in only one of these two sets, not in both simultaneously. The call (message) generation rates of the users of the two sets are λ_v (calls/s/user) for voice users and λ_d (messages/s/user) for data users, both with exponential interarrival time and independent of the satellite to which they connect. At any instant, a user is assumed to be in only one of two states, that is, busy or idle, according to the involvement in a call (data message) transfer. New arrivals are generated only by the idle users, that is, the users that have completed their calls (i.e., have had their data messages acknowledged). A voice call is assumed to originate a continuous bit stream at a constant rate of R_v (Kbps), with an average holding time (exponential distribution) of T_c (s). A data user also is assumed to generate packets from an exponential message length distribution with average length of M kbit. The information is transmitted on a satellite channel with a transmission speed of R_t (Kbps).

In the case of data transmission, packets that fail reception at the destination or are received with uncorrectable errors are not acknowledged; hence, with a random delay, they are retransmitted. Retransmitted packets enter the channel at the rate of λ_r (messages/s). Because the probability of successful transmission is a function of the packet length, the average length of retransmitted messages differs from M for generated messages and has the value of M'; however, its distribution can be assumed to be the same as generated messages, that is, exponential message length. Appropriate selection of retransmission delay in packet CDMA channel using Aloha protocol is an important factor that ensures stability [24]. The equilibrium value of λ_r, the retransmission packet rate, depends on that delay and also on the rate of collision on the channel. At equilibrium, the total packet inflow and outflow rates should be equal. With that fact and with a procedure similar to the one used in [25], the average length of the retransmitted message M' and the retransmitted packet rate λ_r are searched numerically throughout the simulation. Figure 3.12 summarizes the traffic load offered to the channel at the time of process.

To apply the equilibrium condition, it first is necessary to find the probability of packet success. At each satellite and in the absence of thermal noise, the packet error is caused by the interference from all users lying in the interference area of that satellite. At the network analysis level in many spread-spectrum schemes, it is possible to model the channel interference by summing the interference powers and treating the sum as Gaussian noise [26].

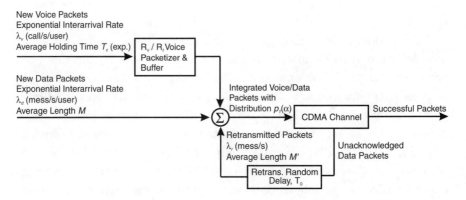

Figure 3.12   Offered traffic load to CDMA channel.

When the interference is assumed as Gaussian noise, we can define the equivalent bit energy-to-noise ratio at the ith satellite, μ_i. By this model, the probability of bit error can be approximated by

$$p_e = 0.5 \, \text{erfc}(\sqrt{\mu_i}) \qquad (3.13)$$

where

$$\text{erfc}(x) = \frac{2}{\sqrt{\pi}} \int_x^\infty e^{-\tau^2} \, d\tau$$

is the complementary error function. The probability of packet success conditioned on μ_i is defined as

$$P[\textit{success of observed packet} \mid \mu_i] = s(\mu_i) \qquad (3.14)$$

In spread-spectrum systems, the function $s(\cdot)$ is a smooth function of signal-to-interference ratio. It depends on the adoption of error-correcting code; with powerful error-correction codes, it approaches a step function at some threshold value of SIR at the satellite. The unconditional packet success probability can be calculated by integrating $s(\cdot)$ with respect to the probability density function of μ_i over all possible values of μ_i.

Because for any value of SIR the probability of packet success and, hence, the expected number of successfully transmitted packets (i.e., throughput) can be evaluated, in this chapter only SIR characteristics as the performance measure of the system are employed. The throughput performance in a more specified system is discussed in Chapter 4.

3.2.3 Simulation environment

To estimate the SIR characteristics of a LEO satellite system in the presence of nonuniform and time-dependent traffic, a simulation model based on the traffic model explained in Section 3.2.2 is used. In this model, a typical LEO satellite system with 11 satellites at the altitude of

1,500 km is assumed. In each processing interval period of the simulation program, T, the traffic uniformity parameter ω changes by the equal steps of $\Delta\omega$ from a maximum value (for nearly uniform traffic load situation) to a minimum value (for a peaked traffic case).

In the simulation model, the multiuser interference power faced by each packet transmitted to the satellite under process is the sum of two parts. The first part is due to the existing packets generated by the users who are in the coverage area of this satellite *and* who select it as a connecting satellite; the second part is the interference from external users in adjacent satellites' coverage areas *and* in line of sight of the satellite under process. Both those interferences are determined as a function of the number of new generated packets, retransmitted packets, and continued packets from last trials, in every trial according to duration time of their connections. Without restricting the discussion to hand-off performance of the system, here we assume that a perfect hand-off procedure for the users has been done; that is, any active user (a user in a busy state) at any instant communicates with the satellite in whose coverage area that user lies that offers minimum required transmitting power to that user. Table 3.1 summarizes the simulation parameters used for evaluation of the performance of our LEO satellite system.

Table 3.1
Simulation Parameters

Item	Symbol	Value
Channel transmission speed (Kbps)	R_t	20
Voice encoder bit rate (Kbps)	R_v	8
Packet size (bits)	L	256
Max. number of correctable errors per packet (bits)	n	10
Synchronization overhead per packet (bits)	—	10
Call generation rate of voice users (calls/s/user)	λ_v	0.0005
Message generation rate of data users (messages/s/user)	λ_d	0.1

Table 3.1 (continued)

Item	Symbol	Value
Average holding time of voice users (minutes)	T_c	3
Average length of message (kbit)	M	1
Retransmission time-out limit (s)	T_o	60
Processing interval time (s)	T	10

Although it is said that the spatial reuse of frequencies and voice activity are important points to increase the capacity of CDMA [11], here we do not explicitly include them in the model. The reason for no consideration of voice activity is that the low bit-rate speech coding under consideration here indirectly exploits short-term burst effects to some degree. On the other hand, the efficiency of spatial reuse in CDMA depends on a number of factors, including the multiplexing efficiency of the CDMA code and the modulation technique employed, which are not necessary to express precisely here for the purpose of our comparison.

3.2.4   Performance measurement

The calculations given in Section 3.1 can be directly used in the case of integrated voice/data traffics, if we change the distribution function of the packets $p(\alpha)$ in (3.8) to $p_1(\alpha)$, which is the composite distribution of the packets transmitted at the time of process by users, including distributions of new generated packets, retransmitted packets, and continued packets. Then, (3.8) in the case of integrated voice/data traffic situation becomes

$$I_i = \int p_1(\alpha) \cdot \min(P_i(\alpha)) \cdot \bar{I_i}^{-2}(\alpha)d\alpha \qquad i = 1, 2, \ldots, N_s \quad (3.15)$$

Similar to Section 3.1, the power of interference at the ith satellite, that is, the DTS, is

$$I_{DTS} = I_i = 2\left[S_i \int_0^{\pi/N_s} p_1(\alpha) \cdot I_i^2(\alpha) \cdot \bar{I}_i^{-2}(\alpha) d\alpha \right.$$

$$\left. + S_{i-1} \int_{\pi/N_s}^{\beta_I} p_1(\alpha) \cdot I_{i-1}^2(\alpha) \cdot \bar{I}_i^{-2}(\alpha) d\alpha \right] \qquad (3.16)$$

where the first term inside the brackets denotes the (half) interference from the users of the DTS, and the second is that from the users of the STS. In (3.16) we have used the symmetry of the model, which results in the factor of 2 in the equation. Similarly, interference at the STS can be found as

$$I_{STS} = I_{i-1} = I_{i+1}$$

$$= S_{i+1} \int_{\pi/N_s}^{3\pi/N_s} p_1(\alpha) d\alpha + S_i \int_{2\pi/N_s - \beta_I}^{\pi/N_s} p_1(\alpha) I_i^2(\alpha) \bar{I}_{i+1}^{-2}(\alpha) d\alpha$$

$$+ S_{i+2} \int_{3\pi/N_s}^{2\pi/N_s + \beta_I} p_1(\alpha) I_{i+2}^2(\alpha) \bar{I}_{i+1}^{-2}(\alpha) d\alpha \qquad (3.17)$$

where the first term is the interference from its own users, and the second and third terms are from the users of the satellites on both sides. It should be noted here that since the effects of the other satellites are out of the interference area of the satellite under process, only the first-order neighboring satellites' users are considered here.

To examine the change in performance of the satellites according to the change in the level of traffic nonuniformity, we change the value of ω as a linear function of time during the simulation. According to Figure 3.3, for our traffic model a value of $\omega = 0.2$ can show a high nonuniformity distribution and a value of $\omega > 5$ can exhibit the uniform distribution. Therefore, in the simulation we consider the change of ω in such a margin. The time duration will be 21 minutes, which is derived according to the following consideration.

According to the law of Newton, explained in Section 1.2, the angular velocity, ω_{vs}, of each satellite can be found from

$$\omega_{vs} = (gm)^{1/2} \cdot r^{-3/2} \tag{3.18}$$

where $(gm)^{1/2} = 631.3482 \text{ km}^{3/2}/\text{s}$; g is gravity constant; m is the mass of the Earth; and r is the radius of the satellite orbit, equal to $R + h$. From (3.18) and simple calculations, at the altitude used in our numerical examples, 1,500 km, the period of a complete rotation of the satellites will be about $T_s = 116$ minutes. Because we assume 11 satellites in each circular orbit, a simulation period equal to 21 minutes, that is, $2 \cdot T_s/N_s$, will be a good trial. This period is equal to the time necessary for a satellite to move above the observed area. Figure 3.13 shows the change of ω according to the time considered in the simulation.

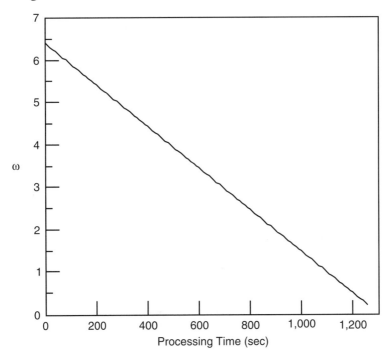

Figure 3.13 Relation between the traffic nonuniformity parameter ω and time during the simulation.

Let us assume that all satellites request the same power levels to the users; thus, $S_i = S$ for $i = 1, 2, \ldots, N_s$. Then the assumption of connecting a given Earth station to the lowest required power satellite results in connection to the nearest satellite from that station. Figure 3.14 shows the simulation result of SIR characteristics at both the DTS and the STS for a minimum elevation angle of $\theta = 10°$, $h = 1,500$ km, and $N_s = 11$ as a function-processing time, assuming equal populations of voice and data users. In Figure 3.14, the simulation procedure starts at $t = 0$ with a large value for ω ($\omega > 5$) as a relatively uniform traffic case and finishes at $t = 1,260$ s with a nonuniform peaked traffic ($\omega = 0.2$). From Figure 3.14, we can find large difference between the signal qualities of the DTS and the STS. At high traffic nonuniformity, the SIR of the DTS degrades notably, while the STS marks superfluous quality. We conclude that the large traffic nonuniformity (e.g., $\omega = 0.2$) decreases the system efficiency, significantly. It should be noted that the traffic nonuniformity

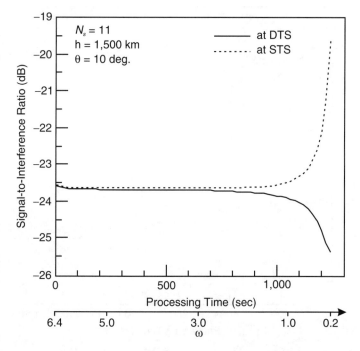

Figure 3.14　SIR characteristics at DTS and STS as a function of traffic nonuniformity for equal populations of voice and data users.

of $\omega = 0.2$ still is not such large nonuniformity; that is, with respect to Figure 3.3, the ratio of traffic under the DTS to traffic of the STS for this value of ω is some value around 10; however, the difference in the SIR performances becomes about 6 dB. Although not shown here, with lower altitudes of satellites, the difference becomes larger (e.g., with $h = 800$ km, the difference increases to 13 dB). That large difference is a direct result of the necessity of connecting users to the nearest satellite. In addition, because for the users located in the coverage area of a single satellite there is no other choice for connecting to the other satellites at a given period of time, they have to accept this large multiuser interference and its consequences.

Figures 3.15 and 3.16 show the SIR characteristics at DTS and STS with the same satellite system constellation parameters as Figure 3.14; but for different ratio of population of data users to voice users. Comparing these three figures, it is concluded that by increasing the ratio of the

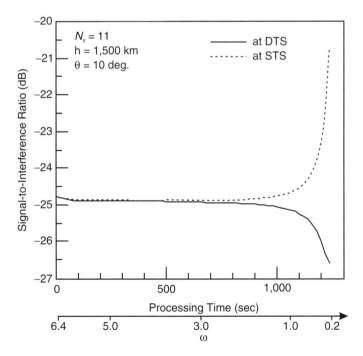

Figure 3.15 SIR characteristics at DTS and STS as a function of traffic nonuniformity for the ratio of data users to voice users equal to 2.

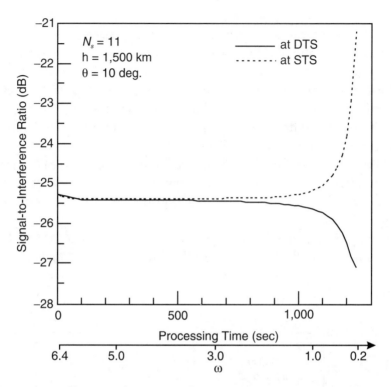

Figure 3.16 SIR characteristics at DTS and STS as a function of traffic nonuniformity for the ratio of data users to voice users equal to 3.

data users to voice users, keeping the total number of users fixed, the performance of the system even in uniform traffic case degrades. One reason is the retransmission permission given to the data users. However, the ratio of data users to voice users seems not to be considerably effective on large divergence in the characteristics of DTS and STS in nonuniform traffic situations.

3.2.5 Dynamic nonuniform traffic concepts

3.2.5.1 Dynamic features of LEO satellite systems
We have analyzed the effect of traffic nonuniformity in LEO satellite systems by defining a nonuniform traffic probability density function, which had a single peak through the coverage areas of three adjacent satellites. In that analysis, it was assumed that at the instant of the process,

the single peak is just under one of the satellites, making its traffic load dense. That assumption leads the analysis to an investigation of a special situation that may have a particular role at the design procedure time. Although LEO satellites are in continuous motion in their orbits and hence their network topology is highly dynamic, the assumption that the system constellation is static for a small period of time is reasonable; in some papers, this situation is referred to as a *quasi-stationary* arrangement of the LEO satellites [27,28].

More exactly speaking, we should note that the traffic loads in coverage areas of LEO satellites are not only nonuniform but also changing as a function of time, as result of two phenomena. The first one is the natural feature of telecommunications systems, that is, the changes in the total number of the users during different hours in a day at the same area, nonuniformity in call arrival and call arrival density, and so on. The second phenomenon is a direct result of the moving property of LEO satellites, from the viewpoint of a fixed object on the Earth. Speaking more precisely, such an object should not be called fixed, when we remember the high speed of the LEO satellites (e.g., with a typical value of $h = 1,500$ km, each LEO satellite has a linear ground speed of about 7.1 km/s), not comparable to the speed of any known vehicles on the ground or in the air. Therefore, any fixed or mobile stations can be viewed as fixed stations in LEO satellite system. Moreover, in the case of a LEO satellite communication system, the latter phenomenon, that is the change in traffic loads according to the movement of satellites, is rapid, compared with the change related to different hours in a day. Hence it is reasonable to consider that rapid change only in the calculations, made the other constant during measurement. That way, we can apply the same LEO satellite system model. Again, to make the effect of traffic nonuniformity clearer and the calculations simpler, only one orbit of the LEO satellite system is considered.

3.2.5.2 Simulation model

To estimate the changes in characteristics of a LEO satellite communications system employing CDMA when the satellites experience nonuniform traffic distribution during their travels, a modified version of the model in Section 3.2.2 is used. In this model, it is assumed that a nonuniform traffic distribution as (3.1) with a predefined value of ω in a specified area within the satellites' path exists. Assuming counterclockwise movement of the satellites, two satellites, namely, the ith satellite and its first neighbor satellite to the right, the $(i + 1)$st satellite, subsequently experience service to the users distributed according to that distribution function. The start point of simulation is when the ith and its first neighbor satellite to the right are in $\alpha = 2\pi/N_s$ and $\alpha = 4\pi/N_s$, respectively, far from the peak of the traffic distribution located at $\alpha = 0$. They rotate in their circular orbit with the constant angular velocity until the ith and its first neighbor satellite to the right reach $\alpha = -2\pi/N_s$ and $\alpha = 0$, respectively. During this period, the ith satellite experiences three specific states of traffic of the STS, the DTS, the STS again, and, of course, their intermediate states, subsequently; however, the first right neighboring satellite before reaching the STS and DTS states starts from a very low traffic state, even less than the STS's state.

Except for the traffic nonuniformity measure, ω, which is fixed in this simulation model, other parameters in Section 3.2.3, including the simulation period of 21 minutes, stand for this simulation too. Figure 3.17 shows the changes of SIR characteristics at the ith satellite and its first neighbor satellite to the right as a function of the processing time for $\omega = 0.2$, assuming equal populations of voice and data users. As the figure illustrates, even in the case of not-so-large traffic nonuniformity as much as $\omega = 0.2$ (compared with $\omega < 0.1$), in not-so-short periods of time large degradation in SIR at the satellites occurs. That is just when the neighbor satellites, whose facilities generally can be accessed by the users to some degree, have large SIR values. If we again assume that the worst situation in system performance is the case in which the signal quality at a satellite has the lowest value, from the figure this is the case where the peak of the traffic load lies just under one of the satellites; that is, the result derived in Section 3.1 is reconfirmed. Figure 3.14 suggests we apply the facilities of the low-traffic neighbor satellites more optimally.

3.2.5.3   Modified power control scheme

The results shown to this point were based on the assumption that all satellites request the same receiving power levels and thus that users connect to the nearest satellite. That means that without paying attention to the number of simultaneous transmissions and the current packet-loss

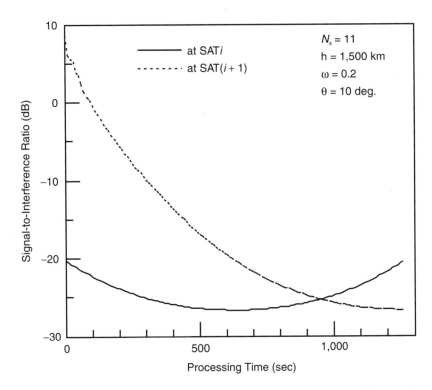

Figure 3.17 Changes in SIR characteristics at the main satellite and its first neighbor satellite to the right as a function of processing time with equal required transmitting power levels.

rate of the system any user must always connect only to the nearest satellite. However, for users located in areas covered by two satellites, there is the choice to connect to the satellite that has a lower traffic load than the closest satellite, even if its distance is larger. This section considers the scheme in which the required uplink power levels to the satellites are changed according to their traffic loads.

In this method, in each processing interval period, T, the traffic load of all satellites distributed in their coverage area, is measured. According to the current value of the required uplink power level to each satellite, its permitted service area and, consequently, the value of SIR at that satellite also are determined. The required uplink power level to any given satellite is changed if the SIR value at it is less than a lower threshold level and at its neighbor satellite is more than an upper threshold level, or vice versa. The change is performed according to the ratio of the traffic loads of that satellite and its neighbors adaptively and is reported to other satellites through the intersatellite network, for example. Because the users located in the coverage area of each satellite and out of its double coverage areas have to connect to that satellite only, the maximum change in required uplink power level to each satellite is limited to the point where that change can affect to the decision of the users in double coverage areas. Any further changes in uplink power levels will only decrease the performance of both light and heavy traffic satellites.

By applying this method, the required uplink power levels to the satellites with heavier traffic load become larger; on the other hand, the required uplink power levels of light-traffic ones decrease. That results in a decrease in the service area of the satellite with the higher traffic load. With the same parameters as in Subsection 3.2.5.2, Figure 3.18 shows

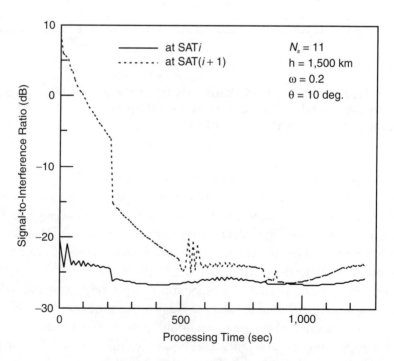

Figure 3.18 Changes in SIR characteristics at the main satellite and its first neighbor satellite to the right as a function of processing time after applying the modified power control scheme.

the simulation results the same as Figure 3.17, here after applying the above explained modified power control method. For this case, it is assumed as an example that the upper and lower threshold levels are −23.0 and −24.5 dB, respectively.

Although the performance improvement due to the modified power control is not well exhibited in Figure 3.18, the method can equalize the traffic load offered to each satellite to some degree. If more satellites exist in each orbit or if their height becomes higher, we can expect that the method gives more performance improvement. Chapter 5 discusses this method in more detail and shows that the method can improve the performance of LEO satellite systems that suffer from geographic traffic nonuniformity.

3.3 Summary

This chapter modeled the situation of nonuniformity in traffic loads of a LEO satellite–based communications system and investigated the performance of the system with a CDMA scheme by the measure of *signal-to-interference ratio* (SIR). Both analog and integrated voice/data traffic scenarios were considered. With the same required uplink power levels requested by the satellites, it was shown that the performance of the LEO satellite system, measured by the value of SIR at each satellite, degrades as a direct result of nonuniformity in distribution of users. It also was shown that the worst case in the performance of the satellite system happens when the peak of the traffic load lies just under one of the satellites.

To have better characteristics near to the case where the traffic distribution is uniform, new traffic assignment and modified power control schemes have been proposed for analog and integrated voice/data systems, respectively, in which weights are given to the required receiving powers of the satellites. By applying those schemes, it has been shown that in nonuniform traffic situations it is possible to improve the performance of the dense-traffic satellite at the expense of degrading the superfluous performance of its neighboring satellites, which have lighter traffic loads. If either the number of satellites on each orbit or the altitude of the satellites becomes higher, we can expect the schemes to exhibit better performances.

References

[1] Gilhousen, K. S., et al., "Increased Capacity Using CDMA for Mobile Satellite Communication," *IEEE J. Select. Areas Commun.*, Vol. 8, No. 4, 1990, pp. 503–514.

[2] Kohno, R., R. Median, and L. B. Milstein, "Spread Spectrum Access Methods for Wireless Communications," *IEEE Commun. Mag.*, Vol. 33, No. 1, 1995, pp. 58–67.

[3] Viterbi, A. J., "A Perspective on the Evaluation of Multiple Access Satellite Communication," *IEEE J. Select. Areas Commun.*, Vol. 10, No. 6, 1992, pp. 980–983.

[4] Viterbi, A. J., "The Evaluation of Digital Wireless Technology From Space Exploration to Personal Communication Services," *IEEE Trans. Vehic. Technol.*, Vol. 43, No. 3, 1994, pp. 638–644.

[5] Abramson, N., "Fundamental of Packet Multiple Access for Satellite Networks," *IEEE J. Select. Areas Commun.*, Vol. 10, No. 2, 1992, pp. 309–316.

[6] Wu, W. W., et al., "Mobile Satellite Communications," *Proc. IEEE*, Vol. 82, No. 9, 1994, pp. 1431–1448.

[7] Jamalipour, A., et al., "A Performance Analysis on the Effects of Traffic Nonuniformity in Low Earth-Orbit Satellite Communication Systems," *Proc. 16th Symp. Information Theory and Its Applications, (SITA' 96)*, Japan, 1993, pp. 203–206.

[8] Jamalipour, A., et al., "Signal-to-Interference Ratio of CDMA in Low Earth-Orbital Satellite Communication Systems With Nonuniform Traffic Distribution," in *Proc. IEEE GLOBECOM '94 Conference*, San Francisco, CA, 1994, pp. 1748–1752.

[9] Jamalipour, A., et al., "Performance of an Integrated Voice/Data System in Nonuniform Traffic Low Earth-Orbit Satellite Communication Systems," *IEEE J. Select. Areas Commun.*, Vol. 13, No. 2, 1995, pp. 465–473.

[10] Jamalipour, A., and A. Ogawa, "Traffic Characteristics of LEOS-Based Global Personal Communications Networks," *IEEE Commun. Mag.*, Vol. 35, No. 2, 1997, pp. 118–122.

[11] Gilhousen, K. S., et al., "On the Capacity of a Cellular CDMA System," *IEEE Trans. Vehic. Technol.*, Vol. 40, No. 2, 1991, pp. 303–312.

[12] Vojcic, B. R., R. L. Pickholtz, and L. B. Milstein, "Performance of DS-CDMA With Imperfect Power Control Operating Over a Low Earth Orbiting Satellite Link," *IEEE J. Select. Areas Commun.*, Vol. 12, No. 4, 1994, pp. 560–567.

[13] Simpson, F., and J. Holtzman, "CDMA Power Control, Interleaving, and Coding," *Proc. 41th IEEE Vehic. Technol. Conf.*, St. Louis, 1991, pp. 362–367.

[14] Shinji, M., *Mobile Communications,* Japan: Maruzen, 1989.

[15] Proakis, J., *Digital Communications, 2nd ed.,* New York: McGraw-Hill, 1989.

[16] Roddy, D., *Satellite Communications,* Englewood Cliffs, NJ: Prentice-Hall, 1989.

[17] Werner, M., et al., "Analysis of System Parameters for LEO/ICO-Satellite Communication Networks," *IEEE J. Select. Areas Commun.,* Vol. 13, No. 2, 1995, pp. 371–381.

[18] Pritchard, W. L., H. G. Suyderhoud, and R. A. Nelson, *Satellite Communication Systems Engineering, 2nd ed,* Englewood Cliffs, NJ: Prentice-Hall, 1987.

[19] Falk, G., et al., "Integration of Voice and Data in the Wideband Packet Satellite Networks," *IEEE J. Select. Areas Commun.,* Vol. SAC-1, No. 6, 1983, pp. 1076–1083.

[20] Ganesh, R., et al., "Performance of Cellular Packet CDMA in an Integrated Voice/Data Network," *Int. J. Wireless Infor. Networks,* Vol. 1, No. 3, 1994, pp. 199–222.

[21] Bertsekas, D., and R. Gallager, *Data Networks,* Englewood Cliffs, NJ: Prentice-Hall, 1987.

[22] Wilson, N. D., et al., "Packet CDMA Versus Dynamic TDMA for Multiple Access in an Integrated Voice/Data PCN," *IEEE J. Select. Areas Commun.,* Vol. 11, No. 6, 1993, pp. 870–884.

[23] Zhang K., and K. Pahlavan, "An Integrated Voice/Data System for Mobile Indoor Radio Networks," *IEEE Trans. Vehic. Technol.,* Vol. 39, No. 1, 1990, pp. 75–82.

[24] Joseph, K., and D. Raychaudhuri, "Stability Analysis of Asynchronous Random Access CDMA Systems," *Proc. IEEE GLOBECOM '86,* 1996, pp. 1740–1746.

[25] Joseph, K., and D. Raychaudhuri, "Throughput of Unslotted Direct-Sequence Spread Spectrum Multiple-Access Channels With Block FEC Coding," *IEEE Trans. Commun.,* Vol. 41, No. 9, 1993, pp. 1373–1378.

[26] Sousa, E. S., and J. A. Silvester, "Optimum Transmission Ranges in a Direct-Sequence Spread-Spectrum Multihop Packet Radio Network," *IEEE J. Select. Areas Commun.,* Vol. 8, No. 5, 1990, pp. 762–771.

[27] Chakraborty, D., "Survivable Communication Concept via Multiple Low Earth-Orbiting Satellites," *IEEE Trans. Aeroso. Electron. Syst.,* Vol. 25, No. 6, 1989, pp. 879–889.

[28] Kaniyil, J., et al., "A Global Message Network Employing Low Earth-Orbiting Satellites," *IEEE J. Select. Areas Commun.,* Vol. 10, No. 2, 1992, pp. 418–427.

4

Spread-Slotted Aloha for LEO Satellite Systems

I N CHAPTER 3, CDMA was assumed as the multiple-access scheme on the uplinks of a LEO satellite system, and the performance of the system under nonuniform traffic situations was examined. In this chapter, we consider, as the second candidate for a multiple-access scheme, a LEO satellite system employing a *direct-sequence spread-slotted Aloha* (DS/SSA) technique and show how the traffic nonuniformity affects the performance of that system.

We start our discussion of employing the DS/SSA scheme in LEO satellite communications networks with an overview of conventional Aloha multiple-access schemes and the idea of spreading Aloha packets to improve the capacity of a system. We review the theory of the spread spectrum and compare the signal forms in CDMA and spread Aloha schemes, to distinguish the two random-access methods. Then, we

117

employ the spread-slotted Aloha in our LEO satellite system and establish a mathematical model for evaluating the performance of the system under nonuniform geographical distribution of users. Finally, we examine the performance of the system under those situations through numerical examples and leave the problem of improving the performance for the following chapters. In such a system, one appropriate measure of the performance is throughput; hence, we use that measure in this chapter. The mathematics given in this chapter is the basic analysis for the following chapters.

For a LEO satellite communication system, spread spectrum is a promising candidate because with advantages such as immunity to interference from adjacent service areas and satellites [1–7]. A packet radio communications system is another candidate, because it is suitable for a system with a huge number of users and relatively small requirements for transmission, as well as a system in which various types of traffic such as voice and data are transmitted at the same time [8–10]. That is one reason why we employ the DS/SSA method, which is a combination of spread-spectrum techniques and slotted Aloha [11–15], as the multiple-access scheme. The discussion of the definition of a LEO satellite system and traffic nonuniformity, however, is not limited to this multiple-access method.

Section 4.2 establishes a new analytical model for a LEO satellite communications system with nonuniform traffic distribution. The model is more useful than the one used in Chapter 3 when packet communications and throughput analysis are involved. In the model presented in this chapter, it is assumed that the location of each user is a random variable and that the users have the same and independent traffic requirements. The results of many conventional studies on traffic analysis can be applied easily to this model. For example, in Section 4.3, the throughput performance of a DS/SSA multiple-access scheme for a LEO satellite system is derived, and the numerical examples shown. It is shown that the performance for each user varies according to the user's location and that the larger possible number of simultaneous transmissions is necessary when nonuniformity of the geographical distribution of the traffic exists. It is also shown that the throughput of the system in a nonuniform traffic assumption is much lower than that of a uniform traffic distribution.

4.1 Spread-slotted Aloha

For more than two decades, the conventional Aloha multiple-access scheme received the attention of not only computer network engineers but also communications network researchers [1,16–22]. That was mostly because of the simplicity offered by the scheme. The Aloha protocol in a packet communications network provides users with nothing but the simplest and most natural feature: Transmit a packet whenever you have one and you want to transmit it. Although this kind of transmission results in large numbers of collisions between simultaneously transmitted packets and low performance, its simplicity is strong enough that even now we see some research on its basic idea. For LEO satellite systems, such a popular scheme should also have promise, even though its basic idea may require some modifications.

4.1.1 The Aloha multiple-access scheme

We start our discussion of spread-slotted Aloha with an overview of the fundamentals of the conventional Aloha scheme. This scheme has a long history with computer and communications engineers.

In the 1970s, Norman Abramson and his colleagues at the University of Hawaii devised a new and elegant method to solve the channel-allocation problem [16]. Since then, their work has been extended by many researchers. Although Abramson's work, called the *Aloha* system, used ground-based radio broadcasting, its basic idea is applicable to any system in which uncoordinated users are competing for the use of a single shared channel.

The basic idea of the Aloha protocol is simple: Transmit your data packet whenever you have data to be sent. That simple protocol, of course, results in collisions, and the colliding packets have to be destroyed. In this system, however, a user can always find out whether a packet was destroyed by monitoring the channel output. In a local area network, that knowledge is immediate; in a geostationary satellite system, for example, it is provided after about a 270-ms delay. If the packet was involved a collision and destroyed, the user waits for a random delay and then retransmits the packet. The latter delay time must be random to avoid collisions of the same packets after retransmission.

Systems in which multiple users share a common channel in a way that can lead to conflicts are widely known as *contention systems*. When a collision between two packets occurs, even if the collision is between the first bit of a new packet and the last bit of a packet almost finished, both packets are totally destroyed, and both have to be retransmitted later. Figure 4.1 illustrates examples of collisions between packets and the transmission of a packet without any collision in an Aloha system. In the figure, equal-length packets are considered.

Now we will look at the efficiency of an Aloha channel. The fraction of transmitted packets that, on average, can pass the channel successfully is referred to as *throughput*. Assume that all packets have the equal length of τ seconds and that the start times of the packets make up a Poisson point process with the parameter λ packets/second [21,23]. With those assumptions, the normalized channel packet rate will be $\xi = \lambda\tau$. A value of $\xi = 1$ corresponds to a channel with packets synchronized perfectly so that the start of one packet always coincides with the end of the previous packet; that is the reason for calling ξ a normalized channel packet rate.

In addition, assume that the start times of the packets plus packet retransmissions make up another Poisson point process. Then we can

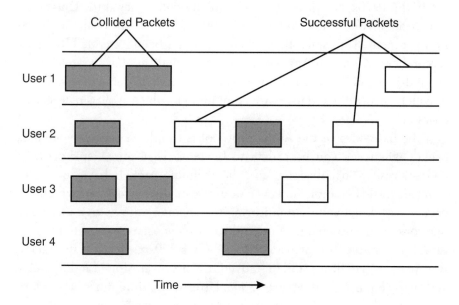

Figure 4.1 Flow of packets on an Aloha channel.

define a quantity G, analogous to the normalized channel packet rate, that takes into account the original packets plus the retransmission packets. The parameter G is called the normalized channel traffic rate, and in general $G \geq \xi$. At low traffic load (i.e., $\xi \approx 0$), there are few collisions, hence few retransmissions, so $G \approx \xi$. At high traffic load, there are many collisions, so $G > \xi$. Under all traffic-load conditions, the throughput is just the offered load, G, times the probability of a transmission being successful, that is, $\xi = GP_0$, where P_0 is the probability that a packet does not suffer a collision.

It easily can be shown that under the above conditions, the relation between the offered traffic load, or G, and the throughput of the Aloha channel is given by

$$\xi = G e^{-2G} \tag{4.1}$$

where e^{-2G} is the probability of no other traffic during the transmission of a given packet in a Poisson distribution scenario. That relationship is plotted in Figure 4.2. As shown in the figure, the maximum throughput occurs at $G = 0.5$, which is about $1/2e$, or 0.184. In other words, the best utilization for an Aloha channel is only about 18%. Of course, that result is not very encouraging, but with users transmitting whenever they want, we cannot expect a higher success rate.

The Aloha method just explained is referred to as *pure Aloha* or *unslotted Aloha*, because there is no synchronization in the start times of packets, so it requires no special control scheme. Just a few years after proposal of the pure Aloha scheme, a simple approach for doubling the capacity of the pure Aloha, called *slotted Aloha*, was proposed [24,25]. In the newer proposal, the time axis was divided into discrete intervals, each corresponding to one packet and called time slots. In a slotted Aloha scheme, a terminal is not permitted to send whenever it has a packet. Instead, it is required to wait for the beginning of the next slot. That control turns the continuous pure Aloha into a discrete one. Therefore, there is no other new traffic generated during a slot, which leads to higher throughput as

$$\xi = Ge^{-G} \tag{4.2}$$

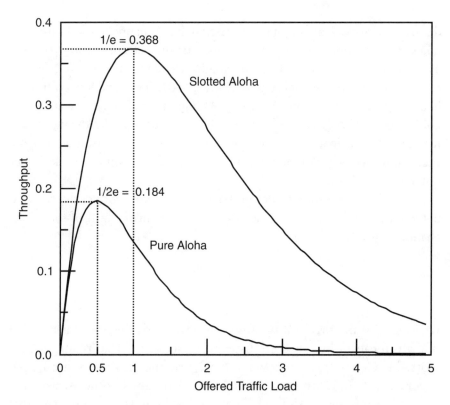

Figure 4.2 Comparison of the performances of two Aloha systems.

The relation between the offered traffic load and the throughput of the slotted Aloha scheme of (4.2) is also plotted in Figure 4.2. As can be seen in the figure, the peak of throughput for the slotted Aloha is enhanced, compared with the pure Aloha, by a factor of 2, and that peak occurs at $G = 1$. That means that if the system is operating at $G = 1$, the probability of an empty slot is $1/e$, or about 0.368. Operating at higher traffic loads reduces the number of "empties" but increases the number of collisions exponentially.

One simply can examine the average delay time for an Aloha channel, that is, the average time required to transmit a given packet, including the retransmissions, and see how that time increases drastically when the offered traffic load exceeds the point that provides maximum throughput of the Aloha channel. A large increase in the average delay time results from the large number of retransmissions for each transmitted packet and the increase in traffic load of the channel. Because that increase affects the stability of system, it should be monitored carefully.

4.1.2   Spreading the Aloha packets

The simplicity offered by the Aloha schemes for a communications network is so important that the method has had the attention of communications engineers for a long time. With the slotted Aloha, even if a kind of condition on transmission of the users is enforced, the basic simplicity of the Aloha scheme is not affected. In slotted Aloha, a user must wait only until the starting point of the next time slot, which is on average half a packet length. However, with this simple approach, the capacity of the channel is enhanced by a factor of 2, and the channel can handle higher traffic loads. In short, with Aloha schemes, random access to a common channel by multiple users is realizable.

The basic idea of an Aloha scheme, even with its low capacity feature, has been employed in some satellite systems [18], which is evidence that the Aloha schemes can match the special features of satellite systems. However, for future satellite systems, the capacity of the Aloha schemes should be improved to handle large amounts of information data.

The problem with the Aloha scheme that limits its capacity is the collisions of the packets. Any approach to enhance the capacity of the Aloha schemes should consider either how the number of collisions can be decreased or how the collisions can be managed so they have less effect on the capacity of the system. Dividing the time axis used in slotted Aloha method limits the number of transmissions during each specific interval or slot. If the number of users in a system is large, collisions are an unavoidable feature, although we may limit the number of collisions. Therefore, for a multiuser system, we should focus on the second consideration, that is, weakening the effect of collisions. (A variety of protocols have been developed to deal with the problem of packets lost due to contention, see [20,21].)

In fact, conventional channel-assignment multiple-access schemes such as TDMA and FDMA also follow the strategy of limiting the number of collisions by preparing a single channel for each user at each time or frequency interval. One also may consider a TDMA or an FDMA system in which users can randomly access each TDMA or FDMA channel with an Aloha-type attempt. That can be considered an application—but not a modification—of the Aloha scheme. By applying the Aloha in TDMA or FDMA, those systems can handle the transmission of users more flexibly, at least, because in Aloha-type TDMA or FDMA systems, we are not limited to a fixed number of times, or frequency slots, and hence a fixed number of users.

In an Aloha system, the probability of packet success is one only when one packet exists; it is zero when two or more packets are on a channel simultaneously. One way to improve the performance of the Aloha channel is to have nonzero packet success probabilities when more than one packet is on the channel. In other words, if we can have a smoother function for the probability of packet success when the number of packets on the channel increases than the simple step function of conventional Aloha, the effect of collisions can be removed. One method to achieve such smooth function is to spread the packets and transmit the spread packets instead of original ones.

Two methods for spreading the packets result in two different modifications to the conventional Aloha scheme. One method is to spread the packets in time, the other is to spread them in frequency. The inventor of Aloha, Norman Abramson, proposed the first method, which he named the *spread Aloha scheme*. The other method is nothing more than the use of spread spectrum and CDMA techniques in conventional Aloha, which can be referred to as *CDMA Aloha*. Although the final mathematical representations for a spread Aloha packet and a CDMA Aloha packet are somehow equivalent, there are some differences between these two schemes, which is the subject of the following subsections. The most significant difference is that in spread Aloha there is only one code, while CDMA Aloha is more similar to conventional CDMA systems. Because the final stage of either scheme has high bandwidth (spread) packets, in much of the literature, CDMA Aloha is also referred to as spread Aloha. In this book, we also consider the CDMA Aloha, that is, a system that uses several orthogonal codes, but we use the name spread Aloha to exhibit the spreading feature of the scheme.

Spread-spectrum multiple-access techniques and Aloha techniques have been studied extensively for satellite and other applications, and their relative merits are well known. Spread Aloha can be assumed to be a combination of the two schemes, in which well-known remarkable features of both schemes are maintained. Moreover, as a particular application of spread spectrum, CDMA systems are being developed as potential candidates for use in digital cellular and satellite mobile radio communications systems. The remainder of this chapter reviews the signaling in the two schemes. Note that the analysis given here is only an introduction to spread-spectrum schemes, necessary for following the studies given in this text. The reader is referred to particular books and papers on spread spectrum for a more complete understanding [26–28].

4.1.2.1 Direct-sequence CDMA signals

This subsection briefly describes the waveforms in a direct-sequence spread-spectrum system, for example, the BPSK modulation. A simple spread-spectrum communication channel is shown in Figure 4.3. As shown in the figure, the information-bearing signal $a(t)$, which is a binary stream of digits, first is modulated by a carrier to produce the signal $s(t)$ such that

$$s(t) = \sqrt{2P} \, a(t) \cos\omega_0 t \qquad (4.3)$$

where P is the power of the signal. After that, the signal $s(t)$ is multiplied by a pseudo-random noise or the spread-spectrum code sequence $g(t)$, which is a binary sequence having the value ± 1 to form the direct-sequence spread-spectrum signal as

$$x(t) = s(t)g(t) = \sqrt{2P} \, a(t)g(t) \cos\omega_0 t \qquad (4.4)$$

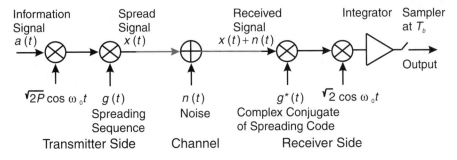

Figure 4.3 An example of a spread-spectrum system.

Here we ignore the characteristics of the spreading code and assume that it is a simple binary sequence. The spread signal is transmitted to the channel and experiences the *additive white Gaussian noise* (AWGN). If the bit rate of the original signal is $f_b = 1/T_b$, where T_b is the time duration of a bit, in a spread-spectrum system the bit rate of the spreading code f_c is selected to be much greater than f_b. Then the multiplication of $s(t)$ by $g(t)$ results in the bits of data being chopped into chips, which is why f_c is called chip rate. The ratio T_b/T_c is referred to as the *processing gain* or simply the gain of the spread-spectrum channel, which shows the bandwidth-expanding factor of the spread-spectrum system and in a practical system has a large value. Because the powers transmitted by $s(t)$ and $x(t)$ are the same, in a spread-spectrum system the power spectral density is reduced by the factor f_b/f_c.

At the receiver side, the direct-sequence spread-spectrum signal is recovered by first multiplying the incoming signal by the complex conjugate of the spread-spectrum code signal and then the carrier $\sqrt{2}\cos\omega_0 t$. Assuming that the spread-spectrum code signal satisfies the condition $|g(t)|^2 = 1$, we have

$$[x(t) + n(t)]g^*(t) = s(t) + n(t)g^*(t) \qquad (4.5)$$

The resulting waveform is then integrated for the bit duration, and the output of the integrator is sampled at the intervals of T_b.

In a CDMA system, each user is given a unique code that is approximately orthogonal (i.e., has low cross-correlation) with the codes of other users. Therefore, the signals of different users can be distinguished from one another with respect to the difference in their codes. Figure 4.4 shows an example of the generating of a CDMA packet. In the figure, each packet contains $n = 6$ bits of information, which is spread by a code of the length of six chips. As can be seen in Figure 4.4, each information bit is replaced by r chips, and the sequence of the chips is according to the code specified for a given user. In a practical situation, however, each packet contains many more bits, and the code length is very long.

After spreading the low-bandwidth information packets with the PN codes, the spread signal has a wide bandwidth that requires r times the bandwidth compared with the original signal. Because each user uses a

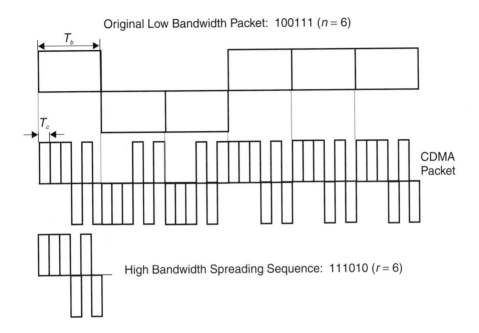

Figure 4.4 An example of generating a CDMA packet.

different code, simultaneous transmissions from other users appear as pseudo-noise to a given user. The packet of a user is recoverable if the power level of noise or multiple-access interference is lower than a specific level. In these systems, different from the conventional Aloha scheme, the probability of packet success is not a (digital) step function, but a smooth function the value of which decreases as the number of simultaneous transmissions increases.

In a CDMA Aloha system, the packets are transmitted to the channel randomly, as in an Aloha system; however, each packet before transmission is spread with a different PN code using CDMA techniques. Therefore, we still have the simplicity of the Aloha, but because the packets are no longer low bandwidth, their information can be recovered even after multiple collisions. Similar to the slotted Aloha system, we can think of a CDMA slotted Aloha system as one in which the packets are transmitted only at the beginning of specific periods of time or slots. That modification requires more complexity in the synchronization process, but it can improve the performance of the system.

4.1.2.2   Spread Aloha signals

Abramson proposed a method to improve the performance of the conventional (unspread) Aloha systems [1]. In his method, the packets are generated directly at a high bandwidth and then are spread in time to make stretched packets. The stretched packets then are multiplied by a spreading sequence, which results in an identical packet, as in the case of CDMA Aloha systems. Figure 4.5 illustrates an example of generating a spread Aloha packet by the Abramson's method.

Comparing the resulting packets in Figure 4.5 with those in Figure 4.4, we can observe that the CDMA packet formed from a spread-spectrum operation is identical to the spread Aloha packet. However, the processes of generating those two packets are different: In the case of CDMA, the packet is formed by multiplying the low-bandwidth

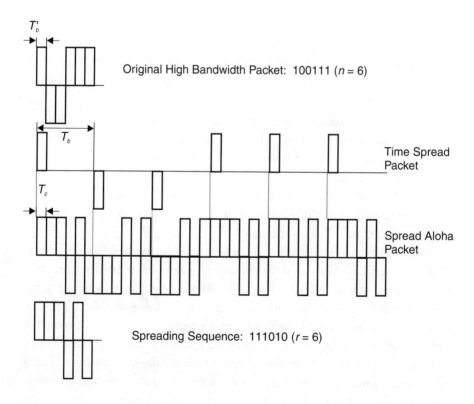

Figure 4.5   An example of generating a spread Aloha packet.

packet by a high-bandwidth carrier, which results in spreading the packet in frequency spectrum. In the case of spread Aloha, however, the packet is generated directly at a high bandwidth, and the packet is then spread in time, which results in an identical relation.

To generate a spread Aloha packet in Abramson's method [1], a kind of transformation is done that can be described in two steps. In the first step, the packets spread in time in a high-bandwidth Aloha channel. The process can be viewed as the dual of the frequency-spreading approach of spread-spectrum multiple access. In the second step, the spread packets pass through a linear filter to replace each information-bearing symbol to spread packets by the spread-spectrum coding sequence. For more details on the mathematics of this method, the reader is referred to [1].

Abramson has explained some differences between the CDMA and his spread Aloha methods [1]. He has mentioned that with his method the energy per information symbol in the original high-bandwidth Aloha packet and that in the spread Aloha packet will be identical. But because in the case of spread Aloha, the same energy is spread over an interval r times longer than in the original packet, the spread Aloha packet can be transmitted by a terminal operating at a much lower transmitter power level. That lower power capability is very important for future personal communication systems with small hand-held terminals.

Abramson also has mentioned that the use of spread spectrum ordinarily implies a nonlinear operation at the transmitter to implement the frequency spreading. The packet in a spread Aloha transmitter, however, originates at the full bandwidth of the channel, and the spreading sequence is integrated into the packet by means of a time-invariant linear operation.

The brief discussion in this section emphasizes the fact that, for future PCNs employing LEO satellites and with large numbers of users equipped with small hand-held digital terminals, the spread Aloha schemes are considerable and promising methods. The selection of the slotted or the unslotted form of the spread Aloha is a synchronization problem in the system, but one that can be solved much easier than the similar one existing in TDMA, for example, when large propagation delays are involved. In the following sections of this chapter, we employ the slotted form of CDMA Aloha, formally referred to as spread-slotted Aloha, in which large enough numbers of codes are prepared to assign to users.

4.2 Employing spread-slotted Aloha in a LEO satellite system

In this section, we employ the spread-slotted Aloha multiple-access scheme in a LEO satellite communications system. As in Chapter 3, we are interested in evaluating the performance of the system in different traffic-load distributions. We are especially concerned with the effect of traffic nonuniformity when such a LEO satellite system is utilized for realizing a global PCN.

For multiuser packet communications systems, there generally are two important measures of performance. The first one is the average number of packets from the total number of transmitted packets that are successfully delivered to the receivers, referred to as throughput. The throughput performance of a packet communications system simply shows how much success a transmitter can expect when sending a packet to a desired receiver. Throughput is very useful in calculating other performance measures of the system. A performance measure closely related to throughput is the probability of packet success expected for a packet transmitting to the channel. Therefore, both the throughput and the probability of packet success are used widely in packet communications systems. In a slotted system, such as slotted Aloha, either in its conventional form or in its spread form, the throughput also can be defined as the expected number of successfully delivered packets in each time slot, with the dimension of packets per slot.

The second measure of performance in a packet communications system is the average delay, defined as the average time elapsed from the moment a packet is generated by a user to the moment the entire packet is received successfully at a satellite. This measure shows, on average, how long a user should expect for delivery of a packet to the receiver. The average delay contains some unavoidable parts, such as the propagation delay that exists physically in any communications systems, and some parts related to the protocol used in the system. Propagation delay in a local area network, for example, is very short and, in general, can be neglected. For a satellite system, however, the length of the propagation delay can be on the order of several packets and, hence, is not negligible. On the other hand, the delay times related to the protocol used in the system should be maintained within acceptable bounds, as far as other parts such as propagation delay, for example, are not avoidable. Generally, no one likes to wait for a long time in a queue to transmit a packet of information.

In fact, in a system that uses retransmission of packets that were not successfully transmitted, such as Aloha systems, long average delay performance means the need for a long interval of time for transmitting each packet. During such long periods, new traffic comes to the channel, and then even normal traffic gradually changes to a high and undeliverable traffic for the network. As a result, the system goes down. This problem often is discussed as a system stability problem. The graphs of throughput and average delay are the complete sets for evaluating the stability in a packet communications network.

Although the throughput and the average delay are two measures of performance in a packet communications system, they are not independent. In fact, they are completely dependent on each other. One reason is that, in a low-throughput system with a large number of users, users expect low probability of success for any packets they transmit on the channel. To be successful in any packet transmission, a user is required to transmit the same packet several times, which in a system with long propagation delay or with long packets means a long average delay in transmission of packets. A system with low throughput performance necessarily has long average delay performance and vice versa. As the first step in the evaluation of the performance of a packet communications system, the measure of throughput seems to be useful enough.

In this chapter, therefore, we consider only the throughput performance of a LEO satellite system. However, because it is important to know how the stability of the system is affected by either an increase in offered traffic load or nonuniformity in traffic, Chapter 6 evaluates the average delay performance as well, after a number of methods that can improve the throughput performance are discussed. It is important for a system engineer to know whether a system fails rapidly or smoothly, and that requires delay performance analysis.

We first develop a mathematical model for distribution of users of our LEO satellite system, which enables us to analyze the throughput performance of this packet communications network. After that, the definitions and calculations on throughput and probability of packet success are presented. The calculations are followed by some numerical examples in Section 4.3.

4.2.1 Distribution of users

Consider a global communication network comprising a number of LEO satellites and a large but finite number of users wanting to communicate with each other through those satellites. In some proposals, the LEO satellite systems are proposed in conjunction with the existing terrestrial systems (e.g., see [29,30]). To analyze the performance of the LEO satellite systems specifically, here we consider a pure LEO satellite system and hence limit the communications of the users through only those satellites.

In this system, the uplink multiple access from users to satellite is established according to DS/SSA techniques. Users start the transmission of their information in the form of the fixed-length packets at common clock instances whenever they have information to send. Packet length and slot size are assumed to be equal. On the downlinks, from the satellites to the users, the satellites can take care of all their transmissions; hence, any conventional multiple-access scheme, such as TDMA, can be applied successfully, making the uplink the limiting direction.

Network topology of the LEO satellite system is highly dynamic; however, a constant topology during each slot interval (i.e., during the transmission of a packet) [31,32] can be assumed; hence, the interference level is assumed to be constant during each time slot. It also is assumed that the interference level for each slot is independent of the other slots. The satellite system model is the same as the model considered in Section 2.3. That is, we assume a multiorbit satellite system with N_s satellites in each circular orbit and a minimum elevation angle of θ_{min}.

As mentioned in the preceding discussions of the performance of the LEO satellite communication systems, it is necessary to consider the effect of nonuniform geographical distribution on the requirements for communications [6,33]. For that reason, this section models the traffic nonuniformity according to a nonuniform probability of existence for each user in a specific area, as described next.

To simplify matters and to show the effect of the traffic nonuniformity clearly, the discussion here focuses on a series of three satellites and their users. Assume that a total number of users, N_u, are distributed randomly in a given area, the size of which is equal to the service areas of three adjacent satellites and their interference areas. For a LEO satellite system in which the satellites have the altitude h (km), the (geometric) interference limit for each of them in a smooth spherical Earth assumption is given by the angle $\beta_I = \cos^{-1}(R/R + h)$, where R is the average radius of the Earth, about 6,378 km. By this notation, we can say that N_u users are distributed in an area whose size equals $4\pi/N_s + 2\beta_I$, which is the size of the interference areas of three succeeding satellites. In that area, it is assumed that the location of every user is a random variable with probability density function $p(\alpha)$ as

$$
p(\alpha) = \begin{cases} \dfrac{A}{\omega} \exp(-\alpha^2/2\omega^2) & \dfrac{-2\pi}{N_s} - \beta_I \leq \alpha \leq \dfrac{2\pi}{N_s} + \beta \\[2ex] 0 & elsewhere \end{cases} \tag{4.6}
$$

where α is the relative location of a user, measured by the angle at the center of the Earth (see Figure 3.2); ω is the parameter representing uniformity in traffic; N_s, as already mentioned, is the number of satellites in each orbit; and A is a factor that makes the total probability of the existence of a user in the area fixed and equal to 1, when changing ω or N_s, equal to

$$
A = \omega / \int_{-2\pi/N_s - \beta_I}^{2\pi/N_s + \beta_I} \exp(-\alpha^2/2\omega^2) d\alpha \tag{4.7}
$$

Note from (4.6) that an infinite value of ω provides uniform traffic distribution, and small values of that parameter establish nonuniform normal-shape distribution of the traffic, similar to the model discussed in Section 3.1.

Finally, it is assumed that each user sends a packet, including newly generated and retransmitted packets, in each time slot with the probability q. This assumption is often used in the literature, for example, in [34]. If we assume M users in the specific area under consideration, the distribution of the number of packets that are sent simultaneously to the channel becomes binomial with the parameters q and M [20], as

$$f(m; M) = \begin{cases} \binom{M}{m} q^m (1-q)^{M-m} & m \le M \\ 0 & m > M \end{cases} \qquad (4.8)$$

where m is the number of transmitted packets.

4.2.2 Throughput analysis

In slotted Aloha systems, steady-state throughput can be defined simply as the expected number of successfully transmitted packets per time slot. In the case of LEO satellite systems with nonuniform traffic distribution, the expected number of users and hence the composite packet arrival rate are different in each service area. Thus, we normalize the throughput for each satellite by the expected number of users in its service area, $E\{N_i\}$, namely, the normalized throughput for the ith satellite, as

$$\xi_{i, \text{norm}} = \frac{\xi_i}{E\{N_i\}} \text{ (packet/slot/user)}, \quad i = 1, 2, \ldots, N_s \qquad (4.9)$$

where ξ_i is the expected number of successfully transmitted packets of the ith satellite in a time slot.

Assume that c_i is the number of successful transmissions for the ith satellite among simultaneous n_i packets transmitted from its service area when m_i packets are sent at that time slot from its interference area. In that case,

$$P\left[c_i = c \mid n_i = n, m_i = m\right] = \binom{n}{c} P^c_{C,i}(n, m) \left[1 - P_{C,i}(n, m)\right]^{n-c} \qquad (4.10)$$

where $P[B]$ denotes the occurrence probability of event B, and $P_{C,i}(n, m)$ is the probability of success for a packet in the presence of n and m packets in the service area and in the interference area of the ith satellite, respectively. The throughput for this case, that is, the expected number of successful transmissions to the ith satellite, is given by

$$\xi_i(n_i, m_i) = E\{c_i \mid n_i, m_i\} \qquad (4.11)$$

The expectation in (4.11) can be calculated by considering all possible situations of n_i packets transmitted to the ith satellite with their probability of a successful transmission, $P_{C,i}(n_i, m_i)$, and probability of an unsuccessful transmission (i.e., the probability of error), $1 - P_{C,i}(n_i, m_i)$, in the form of summation as

$$\xi_{i(n_i, m_i)} = \sum_{c=0}^{n_i} c \binom{n_i}{c} P_{C,i}^c (n_i, m_i) \left[1 - P_{C,i} (n_i, m_i) \right]^{n_i - c} \tag{4.12}$$

Because the summation equals $n_i \cdot P_{C,i}(n, m)$,

$$\xi_i(n_i, m_i) = n_i \cdot P_{C,i} (n_i, m_i) \tag{4.13}$$

Equation (4.13) gives the throughput for the case when n_i packets are transmitted simultaneously from the service area of the ith satellite, and m_i packets are sent at the same time slot from its interference area. To find the total throughput of the ith satellite, we should average (4.13) for all possible values of n_i and m_i. To calculate that average value, we also should consider the probability of n_i and m_i packets from a total of N_u users. Hence, the throughput of the ith satellite, ξ_i, is the expected value of $\xi_i(n_i, m_i)$, that is,

$$\xi_i = E \left\{ \xi_i(n_i, m_i) \right\}$$

$$= \sum_{M=1}^{N_u} P\left[M_i = M \right] \sum_{m=1}^{M} f(m;M) \sum_{n=1}^{m} P\left[n_i = n \mid m_i = m \right] n \, P_{C,i}(n, m)$$

$$\tag{4.14}$$

where M_i is the number of users in the interference area, and N_u, as defined before, is the total number of users in the interference areas of the $(i - 1)$st, ith, and $(i + 1)$st satellites. To understand (4.14), note that the left summation considers the number of users, the middle summation is the simultaneous transmission distribution for each user, and the rightmost summation involves the simultaneous packet throughput for each transmission. From the theory of probability, the probability of M users in the interference area of the ith satellite, for example, equals

$$P\left[M_i = M\right]$$
$$= \binom{N_u}{M}\left[\int_{\beta_i - \beta_I}^{\beta_i + \beta_I} p(\alpha)d\alpha\right]^M \left[1 - \int_{\beta_i - \beta_I}^{\beta_i + \beta_I} p(\alpha)d\alpha\right]^{N_u - M} \quad (4.15)$$

where β_i is the angle between the ith satellite and the peak of the probability density function, $p(\alpha)$. The conditional probability of n packets in the service area of the ith satellite when there is information on existing m packets in the interference area of that satellite also equals

$$P\left[n_i = n \mid m_i = m\right] = \binom{m}{n}\varepsilon^n(1 - \varepsilon)^{m - n} \quad (4.16)$$

where ε is the probability of a single user in the service area when we know that the user is in the interference area of the same satellite and equals

$$\varepsilon = P\left[n_i = 1 \mid m_i = 1\right] = \frac{\int_{\beta_i - \pi/N_s}^{\beta_i + \pi/N_s} p(\alpha)d\alpha}{\int_{\beta_i - \beta_I}^{\beta_i + \beta_I} p(\alpha)d\alpha} \quad (4.17)$$

The remaining problem is to find the probability of success $P_{C,i}(n, m)$.

4.2.3   Probability of packet success

Calculating the probability of packet success in a spread-spectrum packet communications system is a serious problem that has occupied the attention of researchers for a long time [35–48], both for slotted and unslotted systems. In the case of unslotted systems, the problem is more complicated: The level of multiple-access interference over transmission of a packet changes bit by bit because at any instant there is a probability for transmission of a new packet on the channel. In those systems and to make the analysis easier, some studies assumed a *perfect capture* in the analysis of probability of packet success, which means that the transmission of a packet will be successful if merely the preamble of the packet (or sometimes its first bit) is captured successfully. That assumption gives major simplification in mathematics, but the result is not very realistic. Therefore, several studies were carried out with the assumption of *nonperfect capture*.

In slotted systems, the number of interfering transmissions is constant throughout the entire packet, and, hence, the calculation of probability of packet success is much easier than in unslotted systems. Nevertheless, several studies have been performed to approximate the level of multiple-access interference and to make the mathematics simpler. In those studies, the main effort was to evaluate the probability of data bit error without accounting for the bit-to-bit dependence caused by the multiple-access interference. In [42], for example, the upper and lower bounds on the probability of data bit error by constructing the actual density function for the multiple-access interference, given random signature sequences for all users, were derived. In [39], bounds on the bit error probability for deterministic sequences were developed from the convex properties of the error probability function and the characteristic function of the multiple-access interference components.

Although each method has some advantages and disadvantages, Pursley [35] proposed an approximation to the multiple-access interference that has been used for spread-spectrum systems since then [47,48]. In his method, the summation of multiple-access interference and the background Gaussian noise is approximated by a Gaussian noise. Morrow and Lehnert proposed an "improved Gaussian approximation" for the probability of data bit error subsequent to Pursley's method, which was called the "standard Gaussian approximation" [45]. However, according to their analysis, it has been shown that the standard Gaussian approximation of Pursley is close enough to the bounds given by the improved Gaussian approximation when the number of users is large enough, say more than 10. In our analysis, we also use the standard Gaussian approximation, because in general the number of users is large enough.

In the following analysis is a primary assumption that the network operates under perfect power control so the signals from the users in the service area of a given satellite (and connecting to it), say the ith satellite, reach the satellite by equal power level, S_i. The signals from users connecting to adjacent satellites reach the ith satellite with power levels related to their distances to the connected satellite and the ith satellite and the propagation loss factor.

To calculate the packet success probability, we assume that the level of interference is constant over the transmission of a packet. The noise at the receiver is due to interference from other users and to a constant background noise with two-sided power spectral density, $N_0/2$. We denote the equivalent symbol energy-to-noise ratio at the receiver by $E_b/N_{0\text{eff}}$, where $N_{0\text{eff}}/2$ is an equivalent white-noise two-sided power spectral density for the same signal-to-noise ratio at the receiver. If the received signal at the ith satellite has the power S_i, and the interferers have a total power of I_i, the average symbol energy-to-noise ratio at the satellite in the case of the *direct-sequence scheme with BPSK (DS/BPSK)* with rectangular chip pulse is then

$$\mu_i = \frac{2E_b}{N_{0\text{eff}}} = \left(\frac{2I_i}{3LS_i} + \frac{1}{\mu_0} \right)^{-1} \tag{4.18}$$

where μ_0 is the ratio of bit energy-to-power spectral density of the background noise equal to $2E_b/N_0$, and L is the bandwidth-expansion factor (processing gain). The parameter μ_0 also can be defined as the signal-to-noise ratio in the absence of interferers. The equation also can be derived from the results of Pursley [35].

By modeling the total interference over a packet as Gaussian noise in a spread-spectrum system, the probability of packet success becomes a smooth function of signal-to-interference ratio, defined as the conditional probability of packet success as

$$P\left[\textit{success of observed packet} \mid \mu_i \right] = s(\mu_i) \tag{4.19}$$

The unconditional probability of packet success for a packet that is transmitted in the service area of the ith satellite can be determined from

$$P_{C,i}(n, m) = \int_0^\infty s(\mu) f_{\mu_i}(\mu) d\mu$$

$$= \int_0^\infty [1 - F_{\mu_i}(\mu)] s'(\mu) d\mu \qquad (4.20)$$

where $f_{\mu_i}(\mu)$ and $F_{\mu_i}(\mu)$ are the probability density and the probability distribution functions of the random variable μ_i, respectively; and $s'(\mu)$ is the derivative of $s(\mu)$. Note that the distribution of the interference power and thus μ depend on n and m. Also note that the second expression in (4.20) is obtained after an integration by parts. This kind of change in integral is not true in general, though it is true for the random variables involved here.

To determine the probability of packet success in each service area, it is necessary to find the probability distribution of μ_i. However, this necessity can be changed to the determination of the probability distribution function of I_i, $F_{I_i}(y)$ by using the relation in (4.18) as

$$F_{\mu_i}(\mu) = P[\mu_i \le \mu]$$

$$= P\left[\left(\frac{2I_i}{3LS_i} + \frac{1}{\mu_0}\right)^{-1} \le \mu\right]$$

$$= \begin{cases} 1 - F_{I_i}\left[\dfrac{3LS_i}{2}\left(\dfrac{1}{\mu} - \dfrac{1}{\mu_0}\right)\right] & \mu < \mu_0 \\[2ex] 0 & \mu > \mu_0 \end{cases} \qquad (4.21)$$

For the sake of simplicity, let us define

$$K(\mu) \triangleq \frac{3L}{2}\left(\frac{1}{\mu} - \frac{1}{\mu_0}\right) \qquad (4.22)$$

and rewrite (4.21) as

$$F_{\mu_i}(\mu) = \begin{cases} 1 - F_{I_i}\left(K(\mu) \cdot S_i\right) & \mu > \mu_0 \\ \\ 0 & \mu < \mu_0 \end{cases} \qquad (4.23)$$

The parameter $K(\mu)$ denotes the ratio of the power of the interference to the power of the desired signal at each satellite. The importance of this parameter is that it frees us from separate consideration of the three parameters L, μ_0, and μ. We will show later, in Section 4.3, that the value of this parameter evaluated at some threshold level of the signal-to-noise ratio simply provides the number of simultaneous transmissions to each satellite; hence, we call it *multiple-access capability* [7].

In the case of satellite systems, because of the relatively small effect of shadowing and Rayleigh fading, it is reasonable to assume that the radio signal power is attenuated in proportion to the second power of propagation distance [49]. Under that assumption, the required transmitting power level to the ith satellite equals

$$S_{i,\,\mathrm{req}}(\alpha) = C \cdot S_i \cdot l_i^2(\beta_i - \alpha) \quad i = 1, 2, \ldots, N_s$$

$$= C \cdot S_i \cdot \left[a - b \cos(\beta_i - \alpha) \right] \qquad (4.24)$$

where α is the angle between the user and the peak of probability density function $p(\alpha)$; $l_i(\beta_i - \alpha)$ is the distance between the ith satellite and a user in angular position α; $a = h^2 + 2R^2 + 2Rh$; $b = 2R(R + h)$; and C, a constant with the dimension of the inverse of the squared distance, depends on the wavelength of the carrier.

For each satellite; say the ith satellite, let define three separate areas, as shown in Figure 4.6:

- area 1: the intersection area between the service area of the $(i - 1)$st satellite and the visible area of the ith satellite;

- area 2: the service area of the ith satellite;

- area 3: the intersection area between the service area of the $(i + 1)$st satellite and the interference area of the ith satellite.

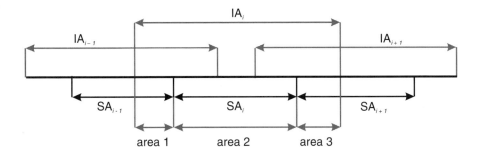

IA: Interference Area
SA: Service Area

Figure 4.6 Explanation of area 1, area 2, and area 3 for the *i*th satellite.

As described in Section 4.2.2, let the number of packets transmitted from area 2 be n and the total number of packets from areas 1, 2, and 3 (i.e., in the interference area) be m. Depending on the location of a given user, different interference power reaches the *i*th satellite during the transmission of the packet transmitted by that user. If we denote the interference power of each signal at the *i*th satellite by $I_{i,\,l}$, where $l = 1$, 2, ..., m, it is equal to

$$
I_{i,l} = \begin{cases}
CS_i - 1 \cdot \dfrac{a - b\,\cos(\beta_i - 1 - \alpha)}{a - b\,\cos(\beta_i - \alpha)} & \text{if transmitted from area 1} \\[2ex]
CS_i & \text{if transmitted from area 2} \\[2ex]
CS_i + 1 \cdot \dfrac{a - b\,\cos(\beta_i + 1 - \alpha)}{a - b\,\cos(\beta_i - \alpha)} & \text{if transmitted from area 3} \\[2ex]
0 & \text{elswhere}
\end{cases}
$$

(4.25)

For each value of interference, y, $F_{I_i}(y)$ equals

$$F_{I_i}(y) = P[I_i \le y]$$

$$= P\left[\sum_{area\ 1\&2\&3} I_{i,l} \le y\right] \qquad (4.26)$$

where the summation is over all m transmitting users in the interference area of the ith satellite. Because, according to (4.25), the interference from n users in the service area does not depend on their location, we can rewrite (4.26) as

$$F_{I_i}(y) = P\left[\sum_{area2} I_{i,1} + \sum_{area\ 1\ \&\ 3} I_{i,1} \le y\right]$$

$$= \begin{cases} P\left[\sum_{area\ 1\ \&\ 3} I_{i,1} \le y - n\ CS_i\right] & \text{if } y \ge n\ CS_i \\ 0 & \text{if } y < n\ CS_i \end{cases} \qquad (4.27)$$

The probability in (4.27) defines the probability of the event that the total interference from the transmitting users in areas 1 and 3 is less than some value at the ith satellite. However, the interferences from the users in those two areas are not necessarily equal, as (4.25) shows. Therefore, to calculate that probability, we can use the following equation:

$$P\left[\sum_{area\ 1\&3} I_{i,1}\ (y - n\ CS_i)\right]$$

$$= \sum_{k=0}^{m-n} P[k \text{ packets from area 1} \mid (m-n) \text{ packets from areas 1\&3}]$$

$$\cdot P\left[I_i(k, m-n) \le (y - nCS_i)\right] \qquad (4.28)$$

where $I_i(k, m-n)$ denotes the sum of interference from the areas 1 and 3 at the ith satellite, when there are k packets from area 1 and $m-n$ packets from areas 1 and 3. The probability of existing k packets from area 1 with the condition that there are $(m - n)$ packets from areas 1 and 3 is

$$
P\left[\begin{array}{c} k \text{ packets from area 1 } | \\ (m - n) \text{ packets from areas 1\&3} \end{array} \right] = \binom{m-n}{k} v^k (1 - v)^{m-n-k}
\tag{4.29}
$$

where v denotes the probability of transmitting from area 1 for a packet equals

$$
v = \int_{area\ 1} p(\alpha)d\alpha \Big/ \left(\int_{area\ 1} p(\alpha)d\alpha + \int_{area\ 3} p(\alpha)d\alpha \right)
\tag{4.30}
$$

To compute $I_i(k, m - n)$ in (4.28), note that this interference power is the sum of the powers of k independent interferences in area 1 and $m - n - k$ independent interferences in area 3. Because these interference powers, $I_{i,\,1}$, have similar source distributions, for enough large values of $m - n$ the summation can be assumed to have a Gaussian distribution with the mean η and the variance σ, expressed as

$$
\eta = k \left[CS_{i-1} \int_{area\ 1} \frac{a - b \cos(\beta_{i-1} - \alpha)}{a - b \cos(\beta_i - \alpha)} p(\alpha)d\alpha \Big/ \int_{area\ 1} p(\alpha)d\alpha \right] + (m - n - k)
$$
$$
\left[CS_{i-1} \int_{area\ 3} \frac{a - b \cos(\beta_{i+1} - \alpha)}{a - b \cos(\beta_i - \alpha)} p(\alpha)d\alpha \Big/ \int_{area\ 3} p(\alpha)d\alpha \right]
\tag{4.31}
$$

$$
\sigma^2 = kC^2 S_{i-1}^2 \left(\left\{ \left[\int_{area\ 1} \left[\frac{a - b \cdot \cos(\beta_{i-1} - \alpha)}{a - b \cdot \cos(\beta_i - \alpha)} \right]^2 p(\alpha)d\alpha \right. \right. \right.
$$
$$
\left. \left. \left. - \left[\int_{area\ 1} \frac{a - b \cos(\beta_{i-1} - \alpha)}{a - b \cdot \cos(\beta_i - \alpha)} p(\alpha)d\alpha \right]^2 \right\} \Big/ \int_{area\ 1} p(\alpha)d\alpha \right)
$$

$$+ (m - n - k)C^2 S_i^2 + 1 \left\{ \left(\left[\int_{area\ 3} \left[\frac{a - b \cdot \cos(\beta_i + 1 - \alpha)}{a - b \cdot \cos(\beta_i - \alpha)} \right]^2 p(\alpha)d\alpha \right. \right. \right.$$

$$\left. \left. \left. - \left[\int_{area\ 3} \frac{a - b \cdot \cos(\beta_i + 1 - \alpha)}{a - b \cdot \cos(\beta_i - \alpha)} p(\alpha)d\alpha \right]^2 \right\} \middle/ \int_{area\ 3} p(\alpha)d\alpha \right)$$

(4.32)

Then, the desired probability can be calculated from [50]

$$P\left[I_i(k, m - n) \le (y - nCS_i) \right] = \frac{1}{2} + \mathrm{erf}\left(\frac{y - nCS_i - \eta}{\sigma} \right) \qquad (4.33)$$

where $\mathrm{erf}(\cdot)$ is the error function, defined as

$$\mathrm{erf}(x) = \left((\frac{1}{\sqrt{2\pi}}) \int_0^x \exp(-y^2/2)dy \right)$$

Note that the constant C will be omitted in the evaluation of (4.33).

4.3 Numerical examples

From the discussion in the preceding section, the probability of success for any given n and m, $P_{C,i}(n, m)$ can be determined if we have $s(\mu)$ defined in (4.19). As an example of determining the function $s(\mu)$, consider a t-error-correcting block code of length n. For any given value of μ, the probability of symbol error is given by [50]

$$P_e = \frac{1}{2} \mathrm{erfc}(\sqrt{\mu}) \qquad (4.34)$$

where

$$\text{erfc}(x) = \frac{2}{\sqrt{\pi}} \int_x^\infty e^{-\tau^2} d\tau$$

is the complementary error function. Then, with the assumption of independent symbol errors, the conditional probability of packet success conditioned on the level of signal-to-noise ratio is given by

$$s(\mu) = \sum_{k=0}^{t} \binom{n}{k} \left[\frac{1}{2} \text{erfc}(\sqrt{\mu}) \right]^k \left[1 - \frac{1}{2} \text{erfc}(\sqrt{\mu}) \right]^{n-k} \tag{4.35}$$

The function $s(\mu)$ depends on the adoption of error-correcting code. With powerful error-correction codes, it is close to a step function at some threshold value of signal-to-noise ratio at the satellite, such as μ_c, as

$$s(\mu) = \begin{cases} 1 & \mu \geq \mu_c \\ 0 & \mu < \mu_c \end{cases} \tag{4.36}$$

As an example, assume that a step function and substitute it into (4.20) with (4.23), we have

$$P_{C,i}(n, m) = \begin{cases} FI_i \left[K(\mu_c) \cdot S_i \right] & \mu > \mu_c \\ \\ 0 & \mu < \mu_c \end{cases} \tag{4.37}$$

Note that $K(\mu_c) + 1$ denotes the possible number of simultaneous transmissions, in the absence of the background noise and interference. Thus, we named $K(\mu_c)$ *multiple-access capability*. Applying that to the discussion in Section 4.2, we can have the throughput characteristics of the DS/SSA in LEO satellite communication systems.

Chapter 3 showed that the worst case in the performance of the LEO satellite communication system with nonuniform traffic distribution occurs when the peak of the traffic load is just under one of the satellites, say, $\beta_i = 0$. This section focuses on that case and shows the results of normalized throughput for two adjacent satellites at the instant where one of the two satellites, say, the ith satellite, is just above the peak of traffic load density. As in Chapter 3, and for simplicity, the ith satellite, which is over the peak of traffic, is the DTS, and its two adjacent neighbor satellites on both sides are the STSs.

Figure 4.7 shows the normalized throughput performances as a function of total offered traffic load, $q \cdot N_u$, for the case where θ_{min} is 10° and $N_s = 11$ satellites on a circular orbit at the altitude of 800 km. The total number of users is assumed to be 100, distributed in the area $4\pi/N_s + 2\beta_l$. Figure 4.7 is for a relatively nonuniform traffic case, $\omega = 0.5$, in which the average number of users of the DTS is about twice

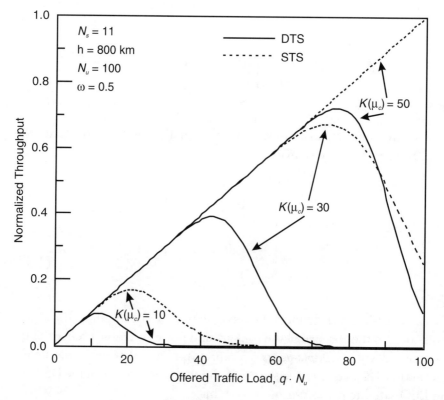

Figure 4.7   Normalized throughput as a function of total traffic load for $\omega = 0.5$.

that of the STS users. From Figure 4.7, two facts can be pointed out. The first observation is the large difference in the performances of the DTS and the STS in the presence of nonuniform traffic. With the same value of $K(\mu_c)$, the performance of a few of the users of the STS is much better than the one for the large number of the users of the DTS. In other words, the traffic nonuniformity alters the quality of service, according to the location of the users and the satellite to which they should be connected.

The second point derived from Figure 4.7 is the performance of the DTS that should serve the major portion of users. As can be seen in the figure, with a large multiple-access capability as $K(\mu_c) = 50$, which is more than the expected number of users of the DTS in this case, in large offered traffic loads the throughput performance of the DTS still decreases. That also can be when $K(\mu_c) = 30$, even though the expected number of the users is much smaller than 30. Because of large interference power reached from the users of DTS, the performance of STS degrades in large offered traffic loads. Therefore, in LEO satellite communication systems, because of the existence of a large interference area compared to service area, the multiple-access capability needs a comparably larger value than the expected number of the users.

Figure 4.8 shows almost the same curves as Figure 4.7, with the same satellite constellation system parameters but for a larger traffic nonuni-formity of $\omega = 0.2$. Note that in this case the expected number of users in the service area of the STS is about one-tenth the number of those in the service area of the DTS. Comparing that figure to Figure 4.7, we observe that with the larger traffic nonuniformity the difference in performance of the DTS and the STS becomes larger. The performance of the DTS for a given $K(\mu_c)$ degrades, while that of the STS improves. Of course, that is because of a decrease in the expected number of users of the STS and an increase in that of the DTS.

To show how the traffic nonuniformity affects the system perform-ance, we evaluate the maximum normalized throughput for the same parameters as for Figure 4.7 and Figure 4.8, that is, the peaks of the curves of those figures, as a function of the traffic uniformity measure, ω, and show the result in Figure 4.9. As expected, under the large traffic nonuniformity, the difference in the performance of the DTS and the STS also is large. When the traffic nonuniformity decreases, that is, ω becomes larger, the performance of the DTS increases and that of the STS

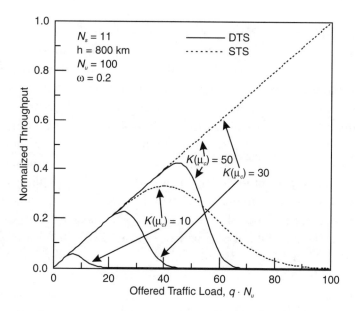

Figure 4.8 Normalized throughput as a function of total traffic load for $\omega = 0.2$.

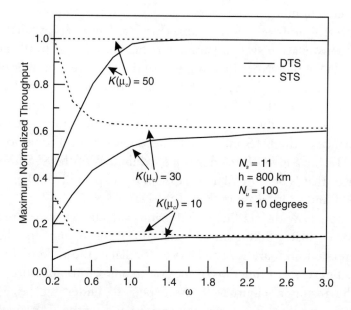

Figure 4.9 Maximum normalized throughput as a function of nonuniformity in traffic for $N_S = 11$ and $K(\mu_C) = 10$, 30, and 50.

decreases. That is caused by the change in the number of the users under each satellite: For STS, it is increased, and for DTS, it is decreased.

Figure 4.10 shows another situation in which we have put more satellites in each orbit. In this case, the area that is located in the coverage area of every two adjacent satellites becomes larger than in the case of Figure 4.9. The figure gives almost the same results: The performance degradation of the STS and improvement for the DTS according to the increase in ω. There is, however, a difference. The performance of the STS increases when ω is large enough for $K(\mu_c) = 30$ and $K(\mu_c) = 50$. The difference is caused by the width of the interference area. In Figure 4.9, the interference areas of two STSs, say, $(i-1)$st and $(i+1)$st satellites, do not have any common area, and a decrease in nonuniformity causes an increase of the number of users in the interference area of the STS and degrades its performance. On the other hand, in Figure 4.10, the interference areas of both STSs overlap each other under the DTS; therefore,

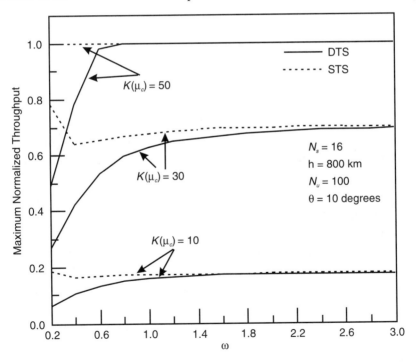

Figure 4.10 Maximum normalized throughput as a function of nonuniformity in traffic for $N_S = 16$ and $K(\mu_c) = 10$, 30, and 50.

the users around the peak of the traffic load located under the DTS affect both STSs at the same time. Thus, the number of users in the overlapping area becomes smaller by making ω larger; the total interference to the STS can be expected to be smaller. Of course, when ω becomes larger, the number of users near the STS also increases and performance degrades. As a result of those two opposite phenomena, as can be seen in Figure 4.10, a dip point appears in the performance of the STS with $K(\mu_c) = 30$.

To show the effect of the traffic nonuniformity on the performance of the system and not only on the performances of the individual satellites, let us define the normalized total throughput as the expected number of successfully transmitted packets in the area that includes the service areas of three succeeding satellites normalized by the total expected number of users in that area, or, in the form of equations as

$$\xi_{norm} = \frac{\xi_{i-1} + \xi_i + \xi_{i+1}}{E\{N_{i-1}\} + E\{N_i\} + E\{N_{i+1}\}} \tag{4.38}$$

Figures 4.11 and 4.12 compare the normalized total throughput characteristics at different degrees of the traffic nonuniformity, with $K(\mu_c) = 30$ and $K(\mu_c) = 10$, respectively. As the traffic nonuniformity increases, the normalized total throughput becomes considerably less than that in the uniform traffic case. That is because of significant degradation in the performance of the satellite above the most probable point, that is, the ith satellite, due to the large number of simultaneously transmitting packets.

It seems from Figure 4.12 that, in large offered traffic load $q \cdot N_u$ with relatively small $K(\mu_c)$ (=10), the performance of a larger nonuniform traffic system such as $\omega = 0.2$ is better than that of a uniform traffic case. That, however, is due to service to a very small portion of users lying in the service areas of the $(i-1)$st and the $(i+1)$st satellites with the expense of large unused parts of their communications facilities. By increasing the multiple-access capability, that phenomenon disappears, that is, in all traffic loads the performance of the uniform traffic case becomes better than that of the nonuniform traffic cases.

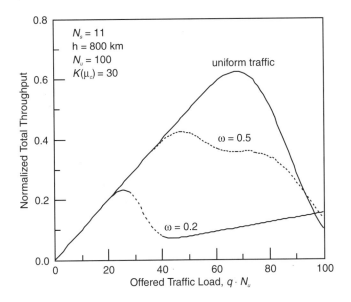

Figure 4.11   Normalized total throughput for different degrees of traffic nonuniformity with $K(\mu_c) = 30$.

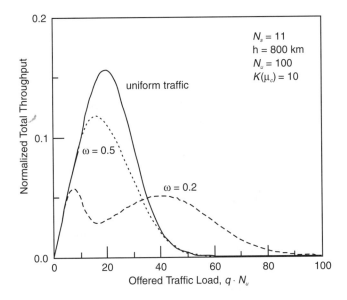

Figure 4.12   Normalized total throughput for different degrees of traffic nonuniformity with $K(\mu c) = 10$.

4.4 Summary

This chapter discussed the fundamentals of the random-access packet communications by the means of LEO satellites. Of importance, we discussed conventional Aloha schemes and explained their simplicity as well as their low capacity. We reviewed the theory of spreading the Aloha packets to reduce the effect of packet collisions and, hence, to improve the performance of the system. We explained the shapes of signals in two multiple-access schemes, the CDMA and the spread Aloha, and discussed the differences between the two when practical specifications are considered, which often are mistakenly considered equivalent schemes.

The second part of this chapter employed the random-access direct-sequence spread-slotted Aloha in a LEO satellite system. We presented the throughput performance calculations for the system, which are used throughout the succeeding chapters. To evaluate the throughput performance, we calculated the probability of packet success and, in that way, introduced a Gaussian approximation for the level of multiple-access interference. The approach used in that calculation is only an example of several methods that were developed for about two decades; hence, the reader is referred to the other approaches used in [35–48]. However, the reader should note that any approach other than the one presented here can be used with the other calculations of throughput of this chapter, if it is replaced by the probability of packet success in the proper equations. For example, in Chapter 6, we consider a fading satellite channel and again calculate the probability of packet success as well as the throughput.

The mathematics given in this chapter were general enough that they can be used in packet communications systems other than the LEO satellite system considered here. As an example of an application of the calculations in this chapter, we evaluated the performance of a LEO satellite system in uniform and nonuniform traffic situations. According to the numerical examples presented in this chapter, it was shown that the traffic nonuniformity gives significant effects to the characteristics of the system and that the throughput of the system in nonuniform traffic situation is much lower than that accessible in a uniform traffic situation, which often is assumed in the literature. Therefore, the analysis and evaluation based on the simplified assumption of uniform traffic distribution are not always correct. Moreover, it was revealed that the interference from users in the service areas of adjacent satellites is one of the main factors that limits the performance of the system. To improve the performance of the systems studied here, multiple spot beams could be used as well as interference-cancellation techniques. The following chapters introduce some methods for remedying the effects of traffic nonuniformity for LEO satellite systems and reducing the multiple-access interference, which can be applied in terrestrial systems too. The performance improvement due to spot-beam antenna on satellites, which is the choice of current big-LEO constellation planners is briefly discussed in Chapter 7.

References

[1] Abramson, N., "Fundamentals of Packet Multiple Access for Satellite Networks," *IEEE J. Select. Areas Commun.*, Vol. 10, No. 2, 1992, pp. 309–316.

[2] Pickholtz, R. L., L. B. Milstein, and D. L. Schilling, "Spread Spectrum for Mobile Communications," *IEEE Trans. Vehic. Technol.*, Vol. 40, No. 2, 1991, pp. 313–322.

[3] Gilhousen, K. S., et al., "Increased Capacity Using CDMA for Mobile Satellite Communication," *IEEE J. Select. Areas Commun.*, Vol. 8, No. 4, 1990, pp. 503–514.

[4] Gilhousen, K. S., et al., "On the Capacity of a Cellular CDMA System," *IEEE Trans. Vehic. Technol.*, Vol. 40, No. 2, 1991, pp. 303–312.

[5] Lee, W. C. Y., "Overview of Cellular CDMA," *IEEE Trans. Vehic. Technol.*, Vol. 40, No. 2, 1991, pp. 291–302.

[6] Jamalipour, A., and A. Ogawa, "Traffic Characteristics of the LEOS-Based Global Personal Communications Networks," *IEEE Commun. Mag.*, Vol. 35, No. 2, 1997, pp. 118–122.

[7] Weber, C. L., G. K. Huth, and B. H. Baston, "Performance Consideration of Code Division Multiple-Access Systems," *IEEE Trans. Vehic. Technol.*, Vol. VT-30, No. 1, 1981, pp. 3–10.

[8] Jacobs, I. M., R. Binder, and E. V. Hoversten, "General Purpose Packet Satellite Networks," *Proc. IEEE*, Vol. 66, No. 11, 1978, pp. 1448–1467.

[9] Tobagi, F. A., "Multiaccess Protocols in Packet Communication Systems," *IEEE Trans. Commun.*, Vol. COM-28, No. 4, 1980, pp. 468–488.

[10] Kahn, R. E., et al., "Advances in Packet Radio Technology," *Proc. IEEE*, Vol. 66, No. 11, 1978, pp. 1468–1496.

[11] Jamalipour, A., et al., "Throughput Analysis of Spread-Slotted Aloha in LEO Satellite Communication Systems With Nonuniform Traffic Distribution," *IEICE Trans. Commun.,* Vol. E78-B, No. 12, 1995, pp. 1657–1665.

[12] Makrakis, D., and K. M. Sundara Murthy, "Spread Slotted ALOHA Techniques for Mobile and Personal Satellite Communication Systems," *IEEE J. Select. Areas Commun.,* Vol. 10, No. 6, 1992, pp. 985–1002.

[13] Polydoros, A., and J. Silvester, "Slotted Random Access Spread-Spectrum Networks: An Analytical Framework," *IEEE J. Select. Areas Commun.,* Vol. 5, No. 6, 1987, pp. 989–1002.

[14] Jamalipour, A., et al., "Spread-Slotted Aloha Throughput in Nonuniform Traffic Situation for LEO Satellite Communication Systems," *Proc. Singapore Int. Conf. Communications Systems (ICCS '94),* Singapore, 1994, pp. 554–558.

[15] Jamalipour, A., et al., "LEO Satellite Communication Systems Under Nonuniform Traffic Distribution With Spread-Slotted Aloha," *Proc. Tech. Report IEICE,* SAT 94-57, Japan, 1994, pp. 15–21.

[16] Abramson, N., "The ALOHA System—Another Alternative for Computer Communications," *Proc. 1970 Fall Joint Comp. Conf.,* 1970, pp. 281–285.

[17] Abramson, N., "The ALOHA System," *Computer Communication Networks,* N. Abramson and F. F. Kuo, ed. Englewood Cliffs, NJ: Prentice-Hall, 1973.

[18] Abramson, N., "Packet Switching With Satellites," *Proc. Natl. Computer Conf.,* Vol. 42, 1973, pp. 695–702.

[19] Abramson, N., "The Throughput of Packet Broadcasting Channels," *IEEE Trans. Commun.,* Vol. COM-25, No. 1, 1977, pp. 233–244.

[20] Kleinrock, L., *Queuing Systems, Vols. 1 and 2,* New York: Wiley, 1975.

[21] Bertsekas, D., and R. Gallager, *Data Networks,* Englewood Cliffs, NJ: Prentice-Hall, 1987.

[22] Tanenbaum, A. S., *Computer Networks, 2nd ed.,* Englewood Cliffs, NJ: Prentice-Hall, 1989.

[23] Papoulis, A., *Probability, Random Variables, and Stochastic Processes,* New York: McGraw-Hill, 1965.

[24] Metzner, J. J., "On Improving Utilization in ALOHA Network," *IEEE Trans. Commun.,* Vol. COM-24, No. 4, 1976, pp. 447–448.

[25] Crozier, S. N., "Sloppy-Slotted ALOHA," *Proc. 2nd Int. Mobile Satellite Conf.,* Ottawa, 1990, pp. 357–362.

[26] Viterbi, A. J., "Spread Spectrum Communications—Myths and Realities," *IEEE Commun. Mag.,* Vol. 17, No. 3, 1979, pp. 219–226.

[27] Viterbi, A. J., "When Not to Spread Spectrum—A Sequel," *IEEE Commun. Mag.,* Vol. 23, No. 4, 1985, pp. 12–17.

[28] Viterbi, A. J., *CDMA: Principles of Spread Spectrum Communication,* Reading, MA: Addison-Wesley, 1995.

[29] Pullman, M. A., K. M. Peterson, and Y. Jan, "Meeting the Challenge of Applying Cellular Concepts to LEO SATCOM Systems," *Proc. IEEE Int. Conf. Communications (ICC '92),* Chicago, IL, Vol. 2, 1992, pp. 770–773.

[30] Chin, L., and J. Chang, "Using Low Earth Orbiting Satellites to Backup the Support of the Existing Ground Mobile Communications," *Proc. IEEE Int. Conf. Communications (ICC '93),* Switzerland, Vol. 3, 1993, pp. 1103–1107.

[31] Chakraborty, D., "Survivable Communication Concept via Multiple Low Earth-Orbiting Satellites," *IEEE Trans. Aeroso. Electron. Syst.,* Vol. 25, No. 6, 1989, pp. 879–889.

[32] Kaniyil, J., et al., "A Global Message Network Employing Low Earth-Orbiting Satellites," *IEEE J. Select. Areas Commun.,* Vol. 10, No. 2, 1992, pp. 418–427.

[33] Jamalipour, A., et al., "Signal-to-Interference Ratio of CDMA in Low Earth-Orbital Satellite Communication Systems With Nonuniform Traffic Distribution," *Proc. IEEE GLOBECOM '94 Conference,* San Francisco, CA, 1994, pp. 1748–1752.

[34] Raychaudhuri, D., "Performance Analysis of Random Access Packet-Switched Code Division Multiple-Access Systems," *IEEE Trans. Commun.,* Vol. COM-29, No. 6, 1981, pp. 895–901.

[35] Pursley, M. B., "Performance Evaluation for Phase-Coded Spread-Spectrum Multiple Access Communication—Part I: System Analysis," *IEEE Trans. Commun.,* Vol. COM-25, No. 8, 1977, pp. 795–799.

[36] Yao, K., "Error Probability of Asynchronous Spread Spectrum Multiple-Access Communication Systems," *IEEE Trans. Commun.,* Vol. COM-25, No. 8, 1977, pp. 803–809.

[37] Borth, D. E., and M. B. Pursley, "Analysis of Direct-Sequence Spread-Spectrum Multiple-Access Communication Over Rician Fading Channels," *IEEE Trans. Commun.,* Vol. COM-27, No. 10, 1979, pp. 1566–1577.

[38] Geraniotis, E. A., and M. B. Pursley, "Error Probability for Direct-Sequence Spread-Spectrum Multiple-Access Communications—Part II: Approximations," *IEEE Trans. Commun.,* Vol. COM-30, No. 5, 1982, pp. 985–995.

[39] Pursley, M. B., D. V. Sarwate, and W. E. Stark, "Error Probability for Direct Sequence Spread-Spectrum Multiple-Access Communications—Part I: Upper and Lower Bounds," *IEEE Trans. Commun.,* Vol. COM-30, No. 5, 1982, pp. 975–984.

[40] Pursley, M. B., "Frequency-Hop Transmission for Satellite Packet Switching and Terrestrial Packet Radio Networks," *IEEE Trans. Inform. Theory,* Vol. IT-32, 1986, pp. 652–667.

[41] Pursley, M. B., and D. Taipale, "Error Probabilities for Spread-Spectrum Packet Radio With Convolutional Codes and Viterbi Decoding," *IEEE Trans. Commun.,* Vol. COM-35, No. 1, 1987, pp. 1–12.

[42] Lehnert, J. S., and M. B. Pursley, "Error Probabilities for Binary Direct-Sequence Spread-Spectrum Communications With Random Signature Sequences," *IEEE Trans. Commun.,* Vol. COM-35, No. 1, 1987, pp. 87–98.

[43] Abdelmonem, A. H., and T. N. Saadawi, "Performance Analysis of Spread Spectrum Packet Radio Network With Channel Load Sensing," *IEEE J. Select. Areas Commun.,* Vol. 7, No. 1, 1989, pp. 161–166.

[44] Storey, J. S., and F. A. Tobagi, "Throughput Performance of an Unslotted Direct Sequence SSMA Packet Radio Network," *IEEE Trans. Commun.,* Vol. 37, No. 8, 1989, pp. 814–823.

[45] Morrow, R. K., and J. S. Lehnert, "Bit-to-Bit Error Dependence in Slotted DS/SSMA Packet Systems With Random Signature Sequences," *IEEE Trans. Commun.,* Vol. 37, No. 10, 1989, pp. 1052–1061.

[46] Yin, M., and V. O. K. Li, "Unslotted CDMA With Fixed Packet Lengths," *IEEE J. Select. Areas Commun.,* Vol. 8, No. 4, 1990, pp. 529–541.

[47] Sousa, E. S., and J. A. Silvester, "Optimum Transmission Ranges in a Direct-Sequence Spread-Spectrum Multihop Packet Radio Network," *IEEE J. Select. Areas Commun.,* Vol. 8, No. 5, 1990, pp. 762–771.

[48] Madhow, U., and M. B. Pursley, "Mathematical Modeling and Performance Analysis for a Two-Stage Acquisition Scheme for Direct-Sequence Spread-Spectrum CDMA," *IEEE Trans. Commun.,* Vol. 43, No. 9, 1995, pp. 2511–2520.

[49] Shinji, M., *Mobile Communications*, Japan: Maruzen 1989.

[50] Proakis, J., *Digital Communications, 2nd ed.*, New York: McGraw-Hill, 1989.

Selected bibliography

"Code Division Multiple Access Networks III," special issue *IEEE J. Select. Areas Commun.,* Vol. 14, No. 8, October 1996.

"Code Division Multiple Access Networks IV," special issue *IEEE J. Select. Areas Commun.,* Vol. 14, No. 9, December 1996.

Glisic, S., and B. Vucetic, *Spread Spectrum CDMA Systems for Wireless Communications,* Norwood, MA: Artech House, 1997.

Prasad, R., *CDMA for Wireless Personal Communications,* Norwood, MA: Artech House, 1996.

5

Modified Power Control in Spread-Slotted Aloha

THE PRECEDING CHAPTERS explained the problem of nonuniformity in geographical distribution of the traffic load in LEO satellite communications systems, its effect on signal quality at the satellites with the measures of the SIR, and the throughput characteristics of the system. It was concluded that traffic nonuniformity considerably affects the performance of the system and thus analysis and evaluation based on the assumption of uniform traffic distribution are not always valid. It also was shown that the performance for each user of a satellite communications system varies according to the user's location. In addition, it was revealed that in LEO satellite systems the interference received from the users in service areas of adjacent satellites is one of the main factors that limits system performance.

One method to decrease the level of interference is to divide the footprints of the satellites into smaller cells, which can be done by the use of spot-beam antennas on satellites. If we consider sharp spots within the footprints, it is possible to limit the interference into each spot area (cell). That would be one of the most serious issues in future big-LEO satellite systems. This issue was explained briefly in Section 2.1 and will be discussed more in Chapter 7. This chapter considers a single-beam satellite system and describes a method to decrease the level of interference. Although the method discussed here is for a single-beam satellite system, it can be considered in a multispot beam system as well.

This chapter, by referring to the LEO satellite system model and the traffic distribution assumptions established in preceding chapters, proposes a new method for remedying the effects of traffic nonuniformity and uses numerical examples to evaluate its performance. The method contains some modifications of the conventional power control scheme used in spread-spectrum systems; hence, we call it the *modified power control scheme.*

The goal of the modified power control scheme is to improve the maximum achievable value of the throughput performance of the system and to make the characteristics of the system as close as possible to the uniform traffic situation. In this scheme, the required uplink powers of the users to their connecting satellites, which are requested by the satellites, are changed. This method has some similarities to the traffic assignment control method proposed in Section 3.1. That section concluded that any improvement in the signal quality at a satellite with heavy offered traffic load (i.e., the DTS) requires large degradation in the signal quality at its neighboring satellites. This chapter shows that the new scheme can improve the throughput characteristics of the DTS while maintaining the level of the throughput of the neighbor satellites, because of the nature of receiving a packet successfully even in relatively low values of signal-to-interference ratio.

Section 5.1 discusses the effect of interference from users located in the service area of adjacent satellites and evaluates the worst case in throughput performance of the LEO satellite systems. Section 5.2 reviews the principles of the conventional power control in spread-spectrum systems and then introduces the concepts of the modified power control scheme and illustrates its features by numerical examples. (For more details on the power control issue, see Section 7.2.) Section 5.2 finishes with some practical notes on the realization of the proposed scheme.

5.1 Worst case in throughput performance

According to the analysis in Chapter 3 of the SIR performance of the LEO satellite systems, the worst case in the performance of the system occurs when a major portion of users are accessing a single satellite simultaneously. From the viewpoint of the nonuniform traffic model established in Chapter 3, that situation occurs when the peak of the traffic load is located under just one of the satellites, because the signal qualities (i.e., the SIR levels) at succeeding satellites have large differences. Chapter 4 briefly assumed that situation as the worst case in performance of the system, analyzed the throughput characteristics, and showed numerical examples. Although from the point of view of the signal-to-noise ratio, that situation is the worst case and although there is a close relation between the throughput and SIR characteristics, it is necessary to confirm that for the throughput the situation is also the worst case.

Another conclusion in Chapters 3 and 4 was on the effects of interferences located outside the service area of a given satellite. It was shown that because all satellites use the same carrier frequency on their uplinks, the interference at any given satellite is the sum of all signal powers that can be reached at that satellite (i.e., the signals of the users in the line of sight of that satellite), regardless of whether the originator of a signal is a user inside or outside the service area of the satellite. Especially in the case where the number of interferers outside the service area of a given satellite is very large, the total interference makes the performance of that satellite much lower than its designed level.

Before introducing the first method for improving the throughput characteristics of our LEO satellite system, this chapter reconfirms the above two conclusions by numerical examples based on the mathematics given in the previous chapter. We first compare the throughput performance in two cases where the interference from outside the service area of the satellites is either considered or ignored. After that, we examine the correctness of the assumed worst case even for throughput performance by considering other probable situations.

5.1.1   Intracell interference versus intercell interference

To perform the comparisons, let us consider again the satellite system and traffic models similar to the ones introduced in Chapter 4. For convenience, the two-dimensional LEO satellite system model and a typical traffic density function at the time when the peak of the traffic load is under just one of the satellites, say, the kth satellite, are repeated in Figure 5.1. Throughout the examples given in this chapter, a typical LEO satellite system with 6 orbits and 11 satellites in each orbit is considered. The orbit height, h, is 800 km, and the minimum elevation angle, θ_{min}, that users can connect to satellites is 10 degrees. The total number of users, N_u, in the area under consideration equals 100. To compare the effect of interferences when their sources are either inside or outside the

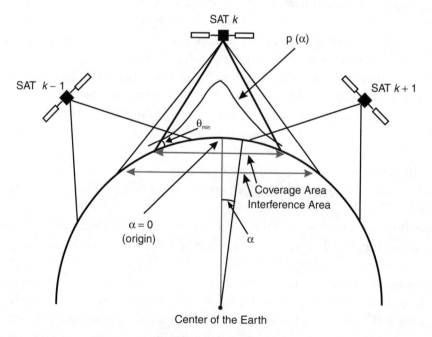

Figure 5.1   Schematic representation of the satellite and traffic model assumptions.

equivalent service area, we use ω, defined in (4.6), as the parameter of traffic uniformity in numerical examples. In most examples, we use typical values of 0.2, 0.5, and 0.75 for ω, which are shown the nonuniform traffic situations, and ∞, which is used for uniform traffic distribution.

Let us evaluate the effect of interference from users outside the service area of satellites on the throughput performance. To perform such an evaluation, we should ignore the interference from users outside the service area of the satellites and compare the results with the ones showed in Chapter 4. (Note that the physical realization of such ignorance requires sharp antenna beams to reject the interference from undesired areas.) Figures 5.2 and 5.3 show the normalized throughput of the *k*th

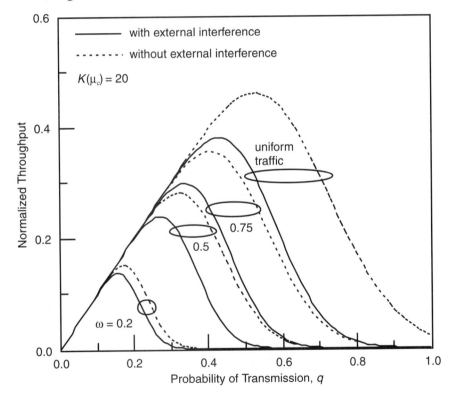

Figure 5.2 Comparison of throughput performances when the interferences from users outside the service area of the *k*th satellite are considered or ignored at different traffic situations of ω equals 0.2, 0.5, 0.75, and ∞ (uniform traffic) with $K(\mu_c) = 20$.

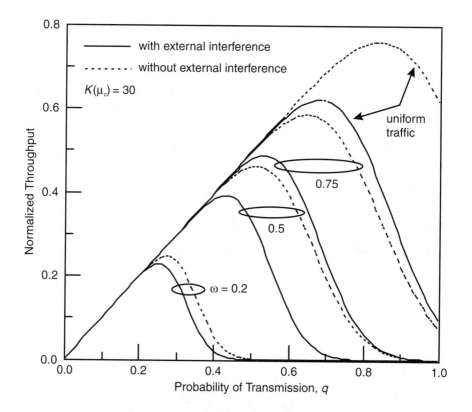

Figure 5.3   Comparison of throughput performances when the interferences from users outside the service area of the *k*th satellite are considered or ignored at different traffic situations of ω equals 0.2, 0.5, 0.75, and ∞ (uniform traffic) with $K(\mu_c) = 30$.

satellite, that is, the satellite over the peak of the traffic load, as a function of composite transmission probability, q; with ω as a parameter, for multiple-access capability; and $K(\mu_c)$ for each satellite equals 20 and 30, respectively. In the figures, the throughput characteristics are compared for the case in which interference from users outside the equivalent service area is considered to the one in which the interference is ignored.

Let us first discuss the throughput performance of the considered satellite in a very high nonuniform traffic distribution, such as $\omega = 0.2$. In that situation, on average about 90% of the users are in the service area of the *k*th satellite; hence, the satellite is faced with a high level of multiple-access interference from the users inside its service area. Even in the situation in which the number of users in the line of sight of the kth satellite but outside its service area is very low, there is considerable difference between the throughput performances when there is and when there is not interference from users outside the service area. For simplicity in discussion, let us adopt the names *intracell interference* and *intercell interference* from terrestrial cellular systems to call the interference from users inside the service area of a given satellite and the interference from users outside the service area of that satellite, respectively. Using those labels, we can say that in high traffic nonuniformity distribution the dominant factor that determines the performance of the satellite with high traffic loads is its intracell interference; however, the effect of intercell interference on the maximum value of throughput also is not negligible.

As the level of nonuniformity in the distribution of users decreases, the effects of intercell interference become larger and larger. At the ultimate, when the traffic distribution is uniform, there is a large-enough difference between the throughput performances in the two considered situations. Therefore, we can say that the effect of interferences from surrounding service areas, that is, the intercell interference, on the throughput performance degradation increases as the traffic distribution of users becomes more uniform. In uniform traffic distribution, the number of users outside the service area reaches its maximum value, and, hence, the largest difference in throughput performances occurs. That means the interferences from users with large distances to their connecting satellites have a large effect on the throughput performance in LEO satellite systems.

The other conclusion drawn from Figures 5.2 and 5.3 is the low throughput in the service area of the considered satellite in the nonuniform traffic situation due to the high level of multiple-access interference. In that situation, most of the time, large numbers of packets are transmitted to a single satellite, and the probabilities of packet success decrease significantly. That result is the same as the one shown in Chapter 4 when the total throughput of the three satellites is considered. In fact, in nonuniform traffic distributions, the dominant factor that determines the total throughout is the throughput of the satellite that services the larger number of users, that is, the kth satellite in the examples given here. We show in Chapter 6 that the traffic nonuniformity not only degrades the throughput performance of the LEO satellite systems but also significantly affects the average delay performance and stability of the system.

5.1.2 Performance of nonworst cases

This subsection discusses the problem of verifying the situation shown in Figure 5.1 as the worst case in throughput performance of the LEO satellite system. Chapter 3 dealt with the same problem but with SIR as the measure of performance. In that chapter, we calculated the values of SIR at two successive satellites during the motion of satellites in their orbit. Considering the instantaneous levels of SIR at those satellites, we found the situation where the values of SIR at those satellites reach the largest difference and then named that situation as the worst case. We have shown that that case occurs just at the instant when one of the satellites is over the peak of the traffic load.

In the case of throughput performance, for which the calculations were shown in Chapter 4, the problem is a bit more complicated. One reason is that in the calculations of the throughput we always use an average value rather than an instantaneous one. As mentioned in Chapter 4, the throughput in a slotted system is defined as the average number of packets successfully transmitted to the channel and received at the satellite in each slot. That means that with the throughput we do not have exact information on the number of packets that successfully exchanged between a given user and the connecting satellite at a specific interval of time. Even if such information is available for every user, it is of no merit from the viewpoint of either individual users or system providers. It is only important for users or system engineers to know how much they can rely on the system when an information packet is transmitted on the channel. Of course, the real statistics of the system simply can be gathered after operation if such information is of interest.

The problem now is how we can analyze and find the worst case in throughput performance. One solution is to average the throughput values (which themselves are averages in nature) for a specific period of time during the motion of satellites and compare those averages for different positions of satellites to the peak of the traffic load. That way is proper when we want to compare the performances of different multiple-access schemes. For example, we will use it in Chapter 6 to compare the conventional spread-slotted Aloha with the scheme that will be proposed in that chapter. However, for evaluation of the worst case, this method is not suitable, because the comparison is made on average values, which are stochastic in nature. In other words, by this method we compare only the averages of some random variables, which sometimes may be provided with not so informative conclusions.

The other way, which engineers usually use, is to compare the situation thought to be the worst case to another critical situation that is expected from general consideration of the system. We use this way, and then we compare the situation shown in Figure 5.1 to another one in which the peak of the traffic load is just in the middle between the two successive satellites, as illustrated in Figure 5.4. In that situation, the angle between the kth satellite and the peak of the traffic load is π / N_s, and, hence, different from the situation of Figure 5.1. In Figure 5.1, the

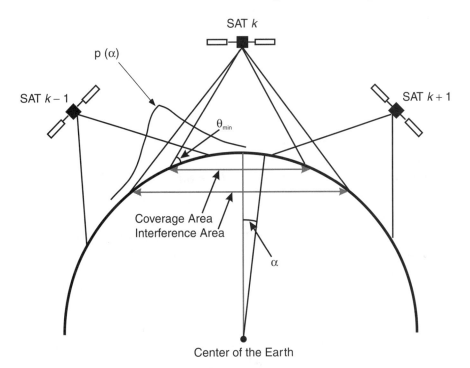

Figure 5.4 Schematic representation of the system model when the peak of the traffic load is located halfway between two successive satellites.

nonuniform traffic distribution of the major portion of users is in the service area of the kth satellite; most users are located in the service areas of two satellites, the kth and the $(k-1)$st ones. The reason for considering this situation for comparison is that any other situation can fall between the two marginal situations in Figure 5.1 and Figure 5.4. It is also worth considering the alternate situation because it happens as often as the Figure 5.1 case.

In the situation shown in Figure 5.1, the major portion of the users should be serviced by a single satellite; hence, the probability of packet success for the packets transmitted in the service area of that satellite is very low. Moreover, the interference caused by those packets to the neighbor satellites makes their performance worse than their designed values. On the other hand, the situation shown in Figure 5.4 is another extreme case, in which the peak of the traffic load is located in the intersection area of the service areas of two neighbor satellites. In the latter situation, the major portion of users can be serviced by two satellites. The comparison of the normalized total throughput, that is, the normalized value of summation of the throughputs of the three satellites, is shown in Figure 5.5. The graph of the situation in Figure 5.1, already referred to as the worst case, is the same as the graph of $\omega = 0.5$ shown in Figure 4.10.

Comparing the two graphs in Figure 5.5, we can see two major differences. First, as may be expected, the peak of the throughput curve for the worst case is lower than that of the other case. In the latter case, the major portion of the users is serviced by two satellites and not only by a single satellite of the worst case. Second, at high offered traffic loads, the curve for the situation in Figure 5.4 has a positive slope. To understand the reason for the latter difference, remember that the total throughput is the sum of the throughputs of succeeding satellites, not just one satellite. Therefore, compared to the worst case, less interference reaches the neighbor satellites, say, the $(k-2)$nd and the $(k+1)$st ones, because the distances between those satellites and the most probable traffic point in the case of Figure 5.4 become longer than those of the worst case. That makes the neighbor satellites have good performances even at high offered traffic loads. The maximum value of the throughput, which usually is referred to as the *capacity* of the system, is an important figure of merit in any packet communications systems. Hence, we con-

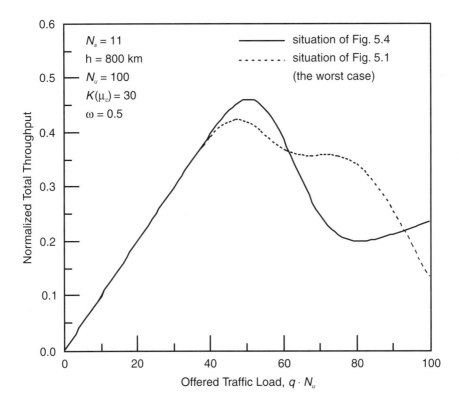

Figure 5.5 Comparison of the normalized total throughput in the two situations in Figures 5.1 and 5.4.

clude that the case shown in Figure 5.1 is the worst case in the performance of the system.

Now we can draw a conclusion on the traffic nonuniformity problem. Because traffic nonuniformity makes the satellites have significantly different traffic loads, use of the communications facilities of the system is not optimal. That is, at a time when some parts of the communications facilities of a satellite are left unused, the high traffic load forced to another satellite makes its characteristics worse than the expected levels. Because the LEO satellites are in continuous motion, the problem is repeated for all satellites whenever they experience areas with high traffic load. As a result, we should seek methods that can make the characteristics of the system closer to a uniform traffic situation. By such a policy, we can expect performance close enough to the designed one for each satellite. In the Section 5.2, such methods applicable in spread-slotted Aloha systems are proposed. Since for a system engineer the important problem is to improve the performance of the system at its worst case, the discussion focuses on the worst case of Figure 5.1.

5.2 Modified power control scheme

Before we discuss the *modified power control* scheme, let us first review the *conventional power control* scheme employed in many multiuser systems, especially in spread-spectrum systems. Of interest are the direct-sequence spread-spectrum systems, in which the transmitter of a desired signal may be located at a greater distance from the receiver and other interfering transmitters situated relatively closer to it. In that situation, the interfering signals could swamp the desired signal if all the transmitters are radiating with equal power. In that case, the signal-to-interference ratio at the receiver is severely degraded, and an increase in the error probability is observed. That phenomenon is known as the *near-far problem*. Under certain circumstances, the near-far problem could become so severe that direct-sequence signaling cannot be used [1]. To achieve an acceptable performance, it is arranged through implementation of power control that all the signals arrive at the receiver with the same average power [2,3]. With all the signals arriving at the receiver with the same average power level, the near-far problem in direct-sequence systems can be eliminated.

The above scenario is the conventional power control scheme, which is realized in different ways, for example, in *open loop* or *closed loop*, in spread-spectrum systems [4–14]. In an open-loop scheme, each user estimates either the total received power from the connecting base station (in our case, the satellite) or the power of a pilot signal sent by that base station. (By monitoring the total received power, rather than using a demodulated signal such as a pilot, measurement can be made rapidly without knowledge of timing, base station identification, or path conditions.) On the basis of on that measure and on a correction supplied by the base station, the users' transmitted powers are adjusted to match the estimated path loss, so as to arrive at the base station at a predetermined level. All users use the same process and attempt to arrive with equal power at the base stations.

In a closed-loop scheme, when uncorrelated differences in the forward and reverse channels cannot be estimated by the users, each user corrects the transmit power with information supplied by the base station on forward links. The base station derives the correction information by monitoring the reverse channel of each user, compares the measurement to a threshold, and requests either an increase or a decrease, depending on the result.

Although there are several parameters, such as channel imperfections and propagation delay, in selection between the two ways of power control, for simplicity in discussion, we assume a perfect open-loop power control in our LEO satellite system and compare the performance of the system utilizing the conventional scheme to the one proposed in Subsection 5.2.1. (The reader is referred to the detailed discussion of power control and system imperfections in a real LEO satellite system given in Chapter 7.)

5.2.1 Purpose and structure of the scheme

In Chapters 3 and 4 and in the preceding section, it was assumed that all transmitted signals from the users are controlled so that they reach their designated satellites at the same level, referred to as perfect power control. The power control is an important necessity in spread-spectrum systems to ensure higher capacity and to avoid the near-far problem, and imperfections in power control result in serious problems in the systems [15–18].

For our LEO satellite system, it was moreover assumed that all signals are received at the same power at all satellites, which means that

$$S_{k-1} = S_k = S_{k+1} = S \qquad (5.1)$$

In addition, it was assumed that each user selects the satellite that requires the lowest transmitting power. Those assumptions realize the equal-size service areas for all the satellites. With the assumption of equal-size service areas of satellites, the actual service area of each satellite has a near hexagonal shape. Also as mentioned before, for each satellite there is a minimum elevation angle that determines an upper bound on the service area of that satellite. The configurations are illustrated in Figure 5.6, in which the instant when the peak of the traffic load is under just one of the satellites is considered.

In the case of a uniform traffic distribution, such a uniform configuration of service areas (i.e., on average, a uniform number of users is provided to each satellite) is natural. In nonuniform traffic situations, however, the throughput of the satellite over the peak of the traffic distribution is the dominant factor in total throughput of the system; thus, we should improve its throughput to enhance the total throughput. One possible way to improve the performance of the satellite is to decrease

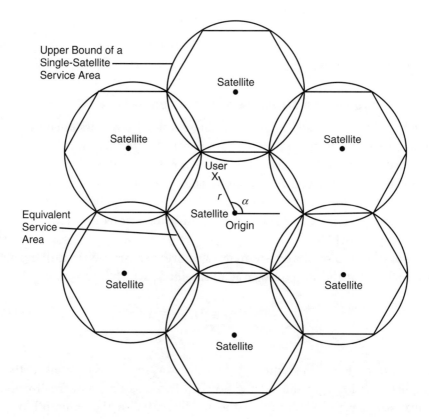

Figure 5.6 Configuration of service areas in the LEO satellite system.

the number of users connecting to it, for example, by decreasing the size of its service area. In a satellite system with overlapping areas between the coverage areas of the adjacent satellites, that decrease can be made by increasing the service areas of the adjacent satellites, without any undesired discontinuity in communications. That means the size of the service areas of the satellites should be adapted according to their traffic loads, at most to the point that the continuity in communication service can be preserved. This is the basic idea of the modified power control scheme, which proposes to make the numbers of the connected users to each satellite closer to the ones of uniform traffic case [19–21]. The method has some similarities with the traffic assignment control method proposed in Section 3.1, here with the measure of the throughput instead of the signal-to-interference ratio.

In the conventional power control scheme, the assumption of connecting users to the satellite that requires the lowest radiated power, with the condition given in (5.1) requires equal-size service areas for all satellites. One approach to realize the idea of the modified power control, that is, changing the size of the service area, is to adapt the condition in (5.1). Note that we still have the assumption of connecting every user to the satellite that requires the lowest transmitting power.

Let us again consider the worst case, where the peak of the traffic load lies under the kth satellite. In that situation, the traffic load of the kth satellite is much higher than those of its adjacent satellites, the $(k + 1)$st and the $(k - 1)$st satellites. Therefore, we adapt the condition in (5.1) to

$$\gamma S_{k-1} = S_k = \gamma S_{k+1} \qquad (5.2)$$

where γ is the ratio of the required receiving powers of the kth satellite to those of its adjacent satellites. In a conventional power control scheme, γ is equal to one; however, in the proposed modified power control scheme, γ has a value larger than or equal to one. If we consider two users, one in the service area of the kth satellite and the other in the service area of the $(k + 1)$st satellite, with equal distances from their connecting satellites, the condition given in (5.2) dictates that the user in the service area of the kth satellite must radiate power higher than the user in the service area of the $(k + 1)$st satellite by the factor γ. According to the value of γ, some users in the overlapping coverage area of two adjacent satellites also find lower required transmitting power to the neighbor satellites; even their distances to the kth satellite are shorter. With large-enough values of γ, it is possible to transfer the total traffic of the overlapping areas to the neighbor satellites and decrease the size of the service area of the kth satellite to the area that can be covered by the kth satellite only. Because the increase in service area of a satellite is limited at most to its upper bound, we should limit the ratio γ to the value where the service areas of the $(k-1)$st and the $(k+1)$st satellites are the same as their coverage areas.

There are different ways to realize the condition given in (5.2). One way is to keep the required receiving powers at the $(k+1)$st and the $(k-1)$st satellites fixed and increase the power S_k by the factor γ. Another way is to keep the S_k constant and decrease the powers S_{k-1} and S_{k+1} by γ. Finally, we can consider an intermediate way: decreasing S_{k-1} and S_{k+1} and increasing S_k so that (5.2) is achieved. Although from the viewpoint of the analysis all three ways result in the same performance, consideration of practical factors, such as limited transmitting power of the hand-held terminals and limited power variations, makes one way more attractive. For example, from the viewpoint of the limitation in the power supply of the terminals, keeping S_k constant and decreasing S_{k-1} and S_{k+1} would be the best. Figure 5.7 illustrates that situation. In the figure, the radiated powers of the users after application of the modified power control scheme are compared with the ones in a conventional power control scheme. As illustrated in Figure 5.7, after employment of the modified power control, the service areas of the $(k-1)$st and $(k+1)$st satellites are expanded and that of the kth satellite is decreased. The radiated power of the users newly included in the increased service areas of the $(k-1)$st and $(k+1)$st satellites also are decreased. That means the total level of the interference at the kth satellite decreases.

5.2.2 Numerical examples

This subsection uses numerical examples to evaluate the performance of the modified power control. The mathematics given in Chapter 4 are used in the calculation involved in these examples with different required receiving powers in (5.2). Also in these examples, a LEO satellite system

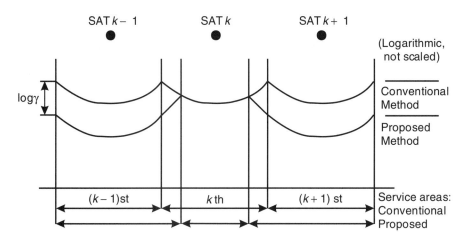

Figure 5.7 Comparison of the radiated powers of the users in conventional and modified power control schemes.

with constellation parameters the same as the ones given in Section 5.1 is assumed. The first example shows the maximum achievable values of the normalized throughputs of the two successive satellites according to the change in γ. The resulting graphs of changes in maximum value of normalized throughput characteristics, that is, the peaks of the normalized throughput curves, for $K(\mu_c) = 50$ as a function of the ratio of the required receiving powers of the satellites are shown in Figure 5.8. In the figure, two cases of traffic nonuniformity for ω = 0.2 and ω = 0.5 are shown. (Note that the modified power control is mostly proposed for remedying the problem of traffic nonuniformity; in a uniform traffic situation, it is better to have a unity value of γ.)

As can be seen in Figure 5.8, by increasing the value of γ from 1, the peak of the throughput performance of the kth satellite can be significantly improved. The interesting point is that the improvement has a very small effect on the degradation in the performance of the $(k + 1)$st satellite if we limit the increase in γ to small values. Therefore, from the viewpoint of the throughput characteristics for each case, we can find a proper value for the ratio of the required receiving powers γ, which makes improvement in the performance of the DTS with negligible degradation in the performance of the sparse traffic one. That result is completely different from the result we had in the case of SIR in Chapter 3, in which we

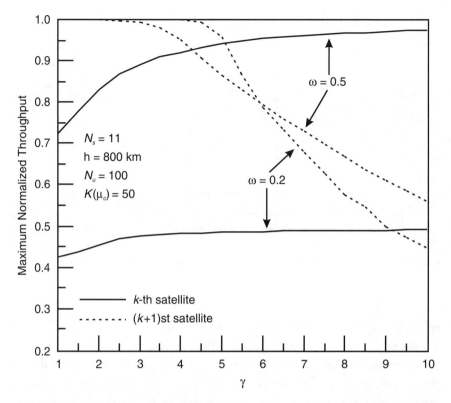

Figure 5.8 Change in the maximum normalized throughput according to the change in γ for $K(\mu_c)$ = 50 and two cases of traffic nonuniformity.

concluded that any improvement in the signal quality of the DTS requires considerable degradation in the performance of its neighbor satellites. In the case of throughput performance, a packet will be successfully received at a satellite if the level of interference is lower than some threshold value. Thus, even in the presence of some degradation in SIR, we still can receive a packet successfully, which is the reason for having the results of Figure 5.8.

In the case of very large traffic nonuniformity such as ω = 0.2, however, the improvement is rather small. The reason is that in that case almost all the users are in the coverage area of the kth satellite, outside the coverage areas of the $(k-1)$st and the $(k+1)$st satellites; for them, there is no other chance to connect to other satellites. The limitation can be made weaker if we supply the satellite system with wider overlapping coverage areas, for example, by having a larger number of the satellites in each orbit or by increasing the altitude of satellites' orbits, which in either case increases the total cost of the system.

Let us now focus on the case of $\omega = 0.5$ and evaluate the performance of the modified power control with a different multiple-access capability of the communications system. Figure 5.9 shows the changes in maximum normalized throughput as a function of the ratio of the required receiving powers for $\omega = 0.5$ and two cases of $K(\mu_c) = 30$ and $K(\mu_c) = 50$. As shown in the figure, with small multiple-access capability such as $K(\mu_c) = 30$, as opposed to the case of $K(\mu_c) = 50$, any improvement in the performance of the satellite with heavy traffic load results in degradation in the performance of its neighbor satellites. We cannot find any value

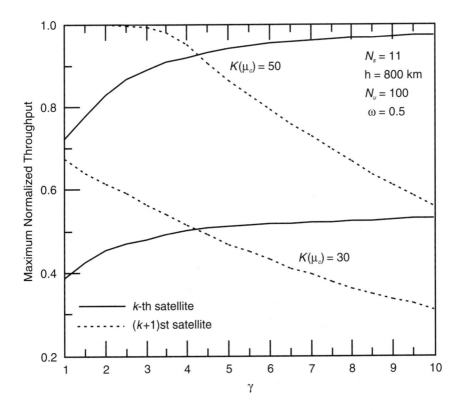

Figure 5.9 Change in the maximum normalized throughput according to the change in γ for $\omega = 0.5$ and different levels of the multiple-access capability, $K(\mu_c) = 30$ and 50.

for the ratio of the required receiving powers that gives the improvement in the performance of one satellite without degradation in the performance of other satellites. The reason is that, with small multiple-access capability, increasing the number of users in each service area degrades the SIR and increases the error probability considerably. However, as will be shown later, even in the case of $K(\mu_c) = 30$, the method still improves the total performance of the system, if the large number of users of the DTS is compared with the ones of its neighbor satellites.

Because the maximum values of the normalized throughput for the kth and its neighbor satellites are achieved at different offered traffic loads, the comparisons made in Figures 5.8 and 5.9 do not give a clear physical meaning. They can, however, suggest proper values for the ratio of the required receiving powers γ for any specific conditions. For example, in Figure 5.9, let us assume a fixed value of $\gamma = 4$ for the case of $K(\mu_c) = 30$ and $\omega = 0.5$. This value is close to the one that makes the peak of the normalized throughput of the kth and the $(k + 1)$st satellites be the same; hence, it may be a proper selection of γ in our modified power control scheme.

The results of the normalized throughput of the kth and the $(k + 1)$st satellites before and after applying the modified power control method with a fixed value of $\gamma = 4$ are shown in Figure 5.10. As stated before, for the case of $K(\mu_c) = 30$ and $\omega = 0.5$, any improvement in the throughput characteristics of the satellite over the dense traffic area results in degradation in the performance of its neighbor satellite, as can be seen in Figure 5.10. However, since the probability of being a user in the service area of the kth satellite in this situation of traffic nonuniformity is higher than being in the service area of the neighboring satellites, total performance improvement of the system can be expected.

To confirm that expectation, let us show the normalized total throughput of three successive satellites. Figure 5.11 shows the normalized total throughput as a function of the offered traffic load for the same parameters as in Figure 5.10. As shown in Figure 5.11, the peak of the throughput performance is improved considerably compared to the case of conventional power control scheme. Moreover, by applying the modified power control, we achieve the throughput characteristics similar in shape to the one appearing in uniform traffic spread-slotted Aloha sys-

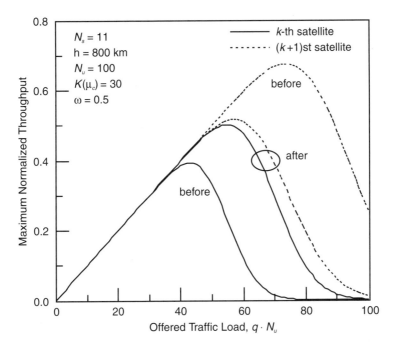

Figure 5.10 Normalized throughputs of the individual satellites for $\omega = 0.5$ and $K(\mu_c) = 30$, before and after applying the modified power control scheme with a fixed value of $\gamma = 4$.

tems. Figure 5.11 illustrates that with the modified power control scheme, even with large traffic nonuniformity, we can realize almost the same characteristics as a uniform traffic distribution, in addition to considerable improvement in throughput performance of the system.

It should be noted that Figure 5.11 is calculated with the assumption of a fixed value for the ratio of powers γ in whole range of the offered traffic load, which is not always the optimum value. The figure suggests that at low offered traffic loads the performance of the system without application of the proposed method is acceptable enough. At heavy offered traffic loads, it also is better to change the ratio of powers to 1 (i.e., disabling the modified power control method). That is, the modified power control is necessary only in a limited range of the offered traffic load, within which the throughput characteristics experience the peak value.

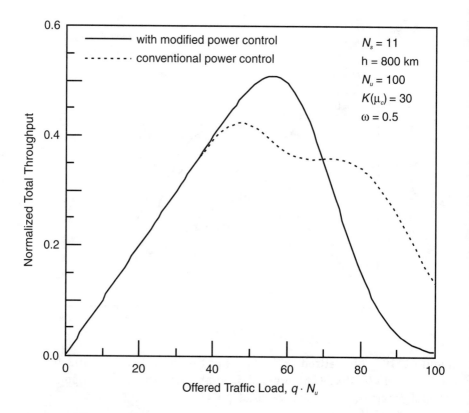

Figure 5.11 Comparison of the normalized total throughput characteristics of the modified power control scheme and the conventional power control, for $K(\mu_c) = 30$ and $\omega = 0.5$, and with a fixed value of $\gamma = 4$.

The performance of such a modified power control scheme with different values of the ratio of powers γ is shown in Figure 5.12. From the figure, we can conclude that, by selecting proper values for the ratio of powers dynamically according to the offered traffic load that gives the best throughput performance for the system, we can expect to have improvement at the whole range of the traffic loads. It should be noted that the method of changing the required receiving powers gives considerable improvement only when the traffic loads of two succeeding satellites have a large-enough difference.

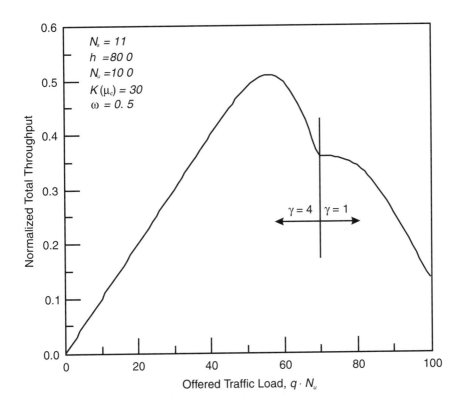

Figure 5.12   Normalized total throughput characteristics by a combination of conventional and modified forms of power control, for $K(\mu_c) = 30$ and $\omega = 0.5$.

5.2.3   Some practical notes on realization of the scheme

This subsection provides some practical considerations on realization of the modified power control scheme in LEO satellite systems. The first point is the selection of the proper value of the ratio of the receiving powers of the satellites, γ. That value can be calculated according to the statistics of the packet transmissions to the satellite channels, for example, by individual satellites. In a LEO satellite network, to realize the modified power control effectively, the satellites should have some communication control channels to each other, which can be provided by intersatellite links, proposed in some global LEO satellite systems. Each satellite can inform other satellites of the statistics of its traffic load via the intersatellite links, then the calculation on γ can be performed either globally for all satellites in a central control station (in space or on the ground) or for a group of satellites. After the calculation, the results on selection of relative values of γ are passed to all satellites, which then can manage the transmitting powers of their users according to the selected value of γ.

Each satellite, according to the updated value of γ, changes the transmitting power of, for example, its pilot signal and sends the control signal over the forward links to all users in its service area. This step is completely executed in the same manner as for the case of conventional power control scheme.

Another point in a realistic system is the range of the offered traffic load in which the system is operating or is designed to operate. The modified power control scheme does not have good performance when the offered traffic load is very high. For that reason, we propose a system in which the value of γ is reset to one to disable the modified power control scheme at heavy offered traffic loads. Actually, in an Aloha or a spread Aloha system, the system is designed to operate at the offered traffic loads near the one that corresponds to the peak of the throughput curves. Operating the system above that point makes the stability of the system sensitive to the changes of the offered traffic load and, hence, lets the system go to unstable situations easily. Theoretically, therefore, there is no need for trying to improve the performance of the system in those areas.

5.3 Summary

This chapter discussed the performance of a LEO satellite communications system in nonuniform traffic situations. By evaluating the performance of one satellite when the effect of multiple-access interference is ignored and comparing the results of that evaluation to the ones shown in Chapter 4, it was shown that the interference from the users in other service areas, referred to as intercell interference, is a dominant factor in degrading the performance of the system. It was also reconfirmed that in nonuniform traffic LEO satellite systems the worst case of the signal-to-interference ratio is the same as the worst case when the throughput characteristics are considered.

According to the numerical examples, it was concluded that the characteristics of the communications system are affected by the presence of traffic nonuniformity and are completely different from the results when the uniform traffic model is considered. It was also concluded that in the presence of the traffic nonuniformity an undesirable behavior of communications system, the dependence of the quality of the service for a user to his location, occurs. To remedy those effects, a modified power control scheme was proposed.

We started the proposal of the method with an overview of the conventional power control schemes used in a spread-spectrum system and discussed different ways of realizing equal receiving powers at the receivers. In the modified power control scheme, the service area of the satellite over the area with heavy traffic load is reduced, and the service areas of its neighbor satellites are expanded. Those changes in the size of the service areas of the satellites are realized by changing the required transmitting powers of the users in each service area, which are requested by the satellites. By changing the power levels, the power of interference also is changed. The result of changing the number of connecting users to each satellite and the total interference power reached at each satellite is improvement of the peak of the total throughput performance of the LEO satellite system. It was also shown that to have better performance over the whole range of the offered traffic loads, it is possible to change the required transmitting powers dynamically. Compared to the results of the traffic-assignment control method, which was proposed in Chapter 3, it was shown that this new method can improve the throughput performance of the LEO satellite system even though the signal qualities at the satellites are affected significantly.

The next chapter introduces other schemes for spread-slotted Aloha that can improve the throughput performance of any multiuser system in heavy traffic situations, including, but not limited to, the nonuniform traffic scenario.

References

[1] Pickholtz, R. L., D. L. Schilling, and L. B. Milstein, "Theory of Spread-Spectrum Communications—A Tutorial," *IEEE Trans. Commun.*, Vol. COM-30, No. 5, 1982, pp. 855–884.

[2] Turin, G. L., "The Effects of Multipath and Fading in the Performance of Direct-Sequence CDMA Systems," *IEEE J. Select. Areas Commun.*, Vol. SAC-2, No. 5, 1984, pp. 597–603.

[3] Kavehrad, M., and P. J. McLane, "Spread Spectrum for Indoor Digital Radio," *IEEE Commun. Mag.*, Vol. 25, No. 6, 1987, pp. 32–40.

[4] Simpson, F., and J. Holtzman, "CDMA Power Control, Interleaving, and Coding," *Proc. 41th IEEE Vehic. Technol. Conf.*, St. Louis, 1991, pp. 362–367.

[5] Gilhousen, K. S., et al., "Increased Capacity Using CDMA for Mobile Satellite Communication," *IEEE J. Select. Areas Commun.*, Vol. 8, No. 4, 1990, pp. 503–514,.

[6] Gilhousen, K. S., et al., "On the Capacity of a Cellular CDMA System," *IEEE Trans. Vehic. Technol.*, Vol. 40, No. 2, 1991, pp. 303–312.

[7] Chuang, J. C.-I., and N. R. Sollenberger, "Uplink Power Control for TDMA Portable Radio Channels," *IEEE Trans. Vehic. Technol.*, Vol. 43, No. 1, 1994, pp. 33–39.

[8] Hanly, S. V., "An Algorithm for Combined Cell-Site Selection and Power Control To Maximize Cellular Spread Spectrum Capacity," *IEEE J. Select. Areas Commun.*, Vol. 13, No. 7, 1995, pp. 1332–1340.

[9] Wong, V., and C. Leung, "A Transmit Power Control Scheme for Improving Performance in a Mobile Packet Radio System," *IEEE Trans. Vehic. Technol.*, Vol. 43, No. 1, 1994, pp. 174–180.

[10] Lee, W. C. Y., "Overview of Cellular CDMA," *IEEE Trans. Vehic. Technol.*, Vol. 40, No. 2, 1991, pp. 291–302.

[11] Pickholtz, R. L., L. B. Milstein, and D. L. Schilling, "Spread Spectrum for Mobile Communications," *IEEE Trans. Vehic. Technol.*, Vol. 40, No. 2, 1991, pp. 313–322.

[12] Kchao, C., and G. L. Stuber, "Analysis of a Direct-Sequence Spread-Spectrum Cellular Radio System," *IEEE Trans. Commun.*, Vol. 41, No. 10, 1993, pp. 1507–1516.

[13] Viterbi, A. M., and A. J. Viterbi, "Erlang Capacity of a Power Controlled CDMA System," *IEEE J. Select. Areas Commun.*, Vol. 11, No. 6, 1993, pp. 892–900.

[14] Sheikh, A., Y. Yao, and S. Cheng, "Throughput Enhancement of Direct-Sequence Spread-Spectrum Packet Radio Networks by Adaptive Power Control," *IEEE Trans. Commun.*, Vol. 42, No. 2/3/4, 1994, pp. 884–890.

[15] Monk, A. M., and L. B. Milstein, "Open-Loop Power Control Error in a Land Mobile Satellite System," *IEEE J. Select. Areas Commun.*, Vol. 13, No. 2, 1995, pp. 205–212.

[16] Vojcic, B. R., R. L. Pickholtz, and L. B. Milstein, "Performance of DS-CDMA With Imperfect Power Control Operating Over a Low Earth Orbiting Satellite Link," *IEEE J. Select. Areas Commun.*, Vol. 12, No. 4, 1994, pp. 560–567.

[17] Newson, P., and M. R. Heath, "The Capacity of a Spread Spectrum CDMA System for Cellular Mobile Radio With Consideration of System Imperfections," *IEEE J. Select. Areas Commun.*, Vol. 12, No. 4, 1994, pp. 673–684.

[18] Cameron, R., and B. Woerner, "Performance Analysis of CDMA With Imperfect Power Control," *IEEE Trans. Commun.*, Vol. 44, No. 7, 1996, pp. 777–781.

[19] Jamalipour, A., et al., "A Modified Power Control Scheme for Remedying the Effects of Traffic Nonuniformity in LEO Satellite Communications Systems," *Int. J. Wireless Information Networks*, Vol. 3, No. 1, 1996, pp. 29–39.

[20] Jamalipour, A., et al., "On Remedying the Traffic Nonuniformity Effects in LEO Satellite Communication Systems," *Proc. 45th IEEE Vehic. Technol. Conf. (VTC '95)*, Chicago, IL, 1995, pp. 986–990.

[21] Jamalipour, A., et al., "LEO Satellite Communication Systems Under Nonuniform Traffic Distribution With Spread-Slotted Aloha," *Proc. Technical Report IEICE*, SAT 94-57, Japan, 1994, pp. 15–21.

6

Transmit Permission Control Scheme for Spread-Slotted Aloha

EMPLOYING SPREAD-SPECTRUM TECHNIQUES in conventional slotted-Aloha multiple-access scheme allows multiple, simultaneous packet transmissions over satellite-based land-mobile communication links [1–3]. With such composite multiple-access schemes, collision of the packets does not result in loss of the information of the packets as long as the SIR level can be kept above some threshold value. In addition, this kind of combination maintains the advantages of both schemes [4–8].

To improve the performance of a communications system employing such a multiple-access scheme, it is necessary to keep the level of multiple-access interference as low as possible. Specifically, if we can keep the level of the multiple-access interference close to the level that the system can support, we can expect to achieve the best throughput performance. One method is to require users to transmit only at the specific situations in which, for example, their distance to the hub station is below a predefined value. In a multicell system with a single hub per cell, this method results in decreasing both the multiple-access interference levels from users inside the cell and from adjacent cells; and hence, improves the performance of the system. That is the basic idea of the *transmit permission control* (TPC) method [9]. Employing such a scheme in cellular systems results in reducing the size of the cells; then, in a fixed communications system, that control results in denial of permission to some portion of the users at all times. Therefore, the TPC method is attractive and practical only in mobile communications systems, where the distance (or, in a fading channel, the propagation loss) between any user and the hub station changes.

This chapter introduces the concept of the TPC scheme and proposes two new adaptive forms of it. In the adaptive schemes, by consideration of the level of the offered traffic loads, permission for the transmission is offered to users more intelligently; therefore, better performance is achieved. This chapter examines, as an example of mobile communications systems, the proposed methods in a LEO satellite communications system.

In this chapter, we first use the throughput analysis of Chapter 4 to evaluate the performance of the new scheme. Moreover, we introduce the mathematics for calculating the average delay performance in LEO satellite systems. The mathematics are used in two steps. First, we show how the average delay performance of the system is affected when a new delay time due to the TPC scheme is added in the system. Next, we consider the effects of nonuniformity in traffic distribution on the average delay or stability of the system. We also modify the throughput analysis for fading channels and discuss the performance of the conventional spread-slotted Aloha and the TPC schemes in the fading channel.

This chapter is organized as follows. Section 6.1 introduces the TPC scheme in nonfading satellite communications channels and evaluates its performance. A performance analysis of the average delay performance also is given in this section. Section 6.2 introduces the fading model and considers the performance of the TPC scheme under this situation. Section 6.3 proposes the concepts of two new adaptive forms of the TPC scheme and compares the performance of those schemes to that of a nonadaptive scheme.

6.1 Transmit permission control scheme: Nonfading channel

This section, like Chapters 4 and 5, is concerned with the performance of a global LEO satellite communications network with the assumption of nonfading or ideal satellite links. Until now, discussion has concentrated on the throughput performance of this packet communications network, which, of course, has a considerable role in any packet communications systems. However, from the viewpoint of stability of the system it is necessary to compute the average delay performance, which has a close relation with the throughput performance. The problem of considering the average delay performance becomes especially important for systems with considerably long propagation delay, such as satellite systems. Even for systems with small propagation delay, such as *local area networks* (LANs), computation of the average delay time is important, because sometimes the number of retransmissions of a packet until successful reception at the receiver becomes very large, and the average delay time reaches unacceptably large values.

In our packet LEO satellite system, both the long propagation delay and retransmission of failed packets exist; hence, calculation of the average delay performance has high-enough importance. It is necessary to remember here that one of advantages of the LEO satellite systems is their low propagation delay compared to geostationary satellite systems; however, compared to the propagation delay in a LAN or terrestrial system, it is long enough to be considered.

Another reason for the importance of average delay performance analysis is to observe how the stability of a system is affected when additional delay times due to the specific protocol utilized in the system are involved. One example of such a case is the protocol that introduced in this chapter, the TPC scheme. In the TPC scheme, to improve the probability of packet success and the throughput performance at some specific intervals, the permission for transmission is partly restrained from users; thus, each user experiences, on average, additional delay. If we show only the throughput performance improvement of such a scheme without discussing its effects on the average delay performance, it will not be a fair comparison with conventional schemes.

6.1.1   Basic considerations

Before starting the main discussion of this section, that is, introducing the TPC scheme, let us do an overview of the basic considerations used in analyses of the throughput and average delay performances of our packet LEO satellite communications network. Therefore, this subsection briefly explains the satellite system model and then provides a new viewpoint on the traffic model to be utilized later.

Consider a global communications network comprising LEO satellites in a multiple-orbit satellite constellation. The total number of orbits and the number of the satellites in each orbit are designed so that at any time any area on the Earth is covered by, at least, one satellite. It is possible for each user of this communications network to communicate with the visible satellite(s), directly.

Because of the relative motion of the LEO satellites, the connection of a user to a satellite is a temporary connection, and a continuous communication must be realized by means of hand-off schemes, mentioned in Section 2.1.2. In this chapter, without delving into the performance of hand-off schemes, we analyze the performance of the system on an average basis during a short period of time in which the mobility of the satellites can be neglected.

The goal of the protocol we propose here is to improve the performance of the communications system on its uplinks, that is, the users-to-satellites direction. On the downlinks, from satellites to users, the satellite, like the base station in a cellular terrestrial system, can take care of all the transmissions in its cell, and any conventional multiple-access scheme such as TDMA can be applied.

To be more specific, assume there are N_s equally spaced satellites on each circular orbit. Each satellite continuously sends out a signal at a constant predefined power level, S. A user, say the jth one, receives at least one attenuated form of this signal, for example from the ith satellite, where $i = 1, 2, \ldots, N_s$, with the power

$$R_{i,\,j} = \kappa \, S \, d_{i,\,j}^{-\beta} \tag{6.1}$$

where κ is a constant with the dimension m^{-2}; $d_{i,j}$ is the distance of the jth user to the ith satellite; and β is the power-loss factor equal to 2 for a satellite link.

The predefined signal is used for three purposes. First, by monitoring this signal, every user can determine the distance to all visible satellites and thus select the nearest one, that is, the satellite that requires the lowest uplink power in a nonfading channel. With the assumption of connecting users to their nearest satellites, equal-size circular service areas for all satellites can be realized. The remaining two purposes are related to the realizations of the power control and the TPC schemes and are described in Subsection 6.1.2.

The remaining problem in describing the system model is defining the spatial distribution of the users. The geographical location of users makes the distribution of the communications traffic loads have different levels of high traffic load peaks over densely populated areas and flat-, or low-, traffic load over unpopulated areas. For a conventional geostationary satellite system, in which every satellite covers about one-third the surface of the globe, the unbalanced traffic load can be managed carefully by each satellite. However, for LEO satellite systems, in which a satellite may cover a limited area of the globe, this problem results in nonoptimal usage of the communication facilities (see Chapters 3 and 4). Although the purpose of this chapter is to introduce a multiple-access method that improves the performance of the system in high traffic situations, not to discuss the traffic nonuniformity problem directly, a spatial distribution model that includes traffic nonuniformity is assumed here. It will be shown that the method can improve the performance of the system in nonuniform traffic distributions as well as in uniform traffic situations.

To make the discussion of performance of the proposed method clear, this chapter considers the performance of one part of the communications system that includes the area covered by three succeeding satellites on the same orbit. It is assumed that the locations of different users are statistically independent. The spatial density of population of the users in this area forms a bell-shaped density function centered at the second satellite. The projection of this bell-shaped density function on the plane that includes the three satellites determines the density of population of users used in the analysis. Note that because of small difference between distances of different users in the same service area to the satellite, this two-dimensional traffic model results in a good approximation with mathematical tractability. Also note that this description is only another viewpoint for the traffic distribution model we have used in the preceding chapters. The total population of users in this area is assumed to be finite and equal to N_u. The location of any user on this plane is assumed to be a random variable with the probability density function

$$p(x) = \frac{A}{\omega} \exp(-x^2/2\omega^2) \qquad (6.2)$$

where x is the relative distance of users from the origin, which is assumed to be under the central satellite; ω is the traffic uniformity parameter; and A is a factor that makes the total probability of existence of a user in this area equal to 1. With this model, we can evaluate the performance improvement achieved by the modification in the multiple-access method in both uniform and nonuniform traffic distributions. That is, a small value of ω realizes a situation in which users are concentrated in the service area of one satellite, which means a high traffic situation for that satellite; enabling us to see the characteristics of the scheme in this case.

6.1.2 Transmit permission control

As stated in Subsection 6.1.1, the performance of the uplinks is of interest. The basic multiple-access scheme assumed in this direction is the *direct-sequence spread-slotted Aloha* (DS/SSA) scheme, which allows multiple transmissions simultaneously and is known as a scheme that increases the capacity of the mobile satellite systems. In conventional DS/SSA systems (those used in Chapters 4 and 5), users transmit information in the form of packets, whenever they have a packet, at the beginning of the next slot, regardless of the status of the channel and the behavior of other users. The purpose of the TPC method is to modify the DS/SSA scheme in such a way that permissions for transmissions are given only to the users whose interference have little effect on the *capture* (success) probability of other packets.

In multiple-access methods based on spread-spectrum techniques, including the DS/SSA scheme, the equalization of the received powers from users is an important issue [10,11]. In LEO satellite systems, any user can, for example, by monitoring the predefined signal received from the nearest satellite, send a packet with proper power level. (Note that in a real communications system, the predefined or pilot signal on the downlink experiences different fading variations to that of the transmitting signal from a user on the uplink. Moreover, the fading on the uplink usually is uncorrelated with that on the downlink, since the uplink and downlink frequency bands usually are separated by more than the coherence bandwidth of the channel. These factors should be considered in power control and are discussed in Section 7.3.) Therefore, it is assumed that all received signals at any given satellite, say, the *i*th one, from the users in its service area arrive at the satellite at the same power level, S. (This is the second usage of the predefined signal mentioned in Subsection 6.1.1.) Therefore, according to (6.1), the transmitted power level of the *j*th user, where $j = 1, 2, \ldots, N_u$, to the *i*th satellite will be

$$T_{i,j} = \kappa^{-1} S d_{i,j}^{\beta} \qquad (6.3)$$

Hence, it is assumed that the network operates under perfect power control. It should be noted that, from the viewpoint of a given packet, all other signals transmitted from users either in the same service area or in different ones act as interference. The packets of users in the same service area have the same power, S; however, the ones from different service areas have different power levels, depending on their distances.

In such a system, the probability of packet success or capture probability for a given packet decreases as the number of simultaneous transmitted packets increases. For example, assume a given user is in service area of the *i*th satellite. As shown in Section 5.1, for a packet transmitted by this user—the tagged packet—the interference from other simultaneously transmitted packets in the same service area and also from transmitted packets in marginal areas of the adjacent service area (because of the relatively long distances of the users in marginal areas to their connecting satellites) are the dominant factors that decrease its capture probability. To improve the capture probability of this typical packet transmitted in the service area of the ith satellite, which results in enhancing the throughput of the system, one possible method is to limit the transmission of the packets from users far from their connecting satellites.

This method can be considered an interference-cancellation method; however, it has a fundamental difference with conventional interference-cancellation schemes used in spread-spectrum systems. In conventional schemes, the existence of a certain level of multiple-access interference is assumed, and the hardware of the receivers as well as the transmitters are redesigned so that the multiple-access interference can be canceled as much as possible. However, in our control of the transmission of packets, we want to decrease the certain level of multiple-access interference while keeping the hardware the same as before by adding some procedures in the processors of the transmitters only. These procedures are accomplished with the existing information at the terminals without additional necessary information.

The idea of the proposed TPC is nothing more than the following simple instruction: Users can send packets if they are located at a distance less than a certain value. Without specifying what determines that distance value at the moment, we can conclude that this method reduces the number of simultaneous transmissions; that is, the level of multiple-access interference, especially by avoidance of the transmissions by users in marginal service areas.

Realization of the method can be performed by the mean of the predefined signal sent by the satellites mentioned in Subsection 6.1.1, as its third usage. Because the powers of the signals are known by all users, any user can determine the propagation loss from a given location to the visible satellites. Then the transmit permission protocol states that a user who has propagation loss larger than a predefined value, such as l_{max}, is not allowed to transmit packets. That means that the users in marginal belts of service areas are prohibited from transmitting their packets.

This scheme may seem unfair to users with relatively large values of propagation loss to their nearest satellites. However, if we allow all users to transmit at all times, the large interference reduces the capture probability and consequently degrades the total throughput of the system. Larger success probabilities for permitted users result in higher total throughput than low success probabilities for all users, which of course depend on proper design of the predefined value of propagation loss, l_{max}. On the other hand, because satellites in LEO satellite systems are in a continuous motion, nonpermitted users are allowed to transmit after a short period, as will be shown later.

Let us have a physical representation of the TPC method. Assume a fixed and same value for the maximum allowable propagation loss, l_{max}, in the service area of all satellites. After applying the method, the radius of service area of each satellite, r_0, is multiplied by some factor ρ, where $0 \le \rho \le 1$, resulting in a reduction of $1 - \rho$ in the radius of service areas. The parameter ρ depends on and has the same information as l_{max} with simpler tractable meaning, that is, large values of l_{max} make the decrement of service area of satellites small, which means a nearly unity value of ρ and vice versa. Figure 6.1 illustrates the change in service areas of the satellites before and after application of the TPC scheme. In the figure, it is assumed that minimum numbers of satellites and orbits are considered for global coverage.

In each circular service area of the satellites in Figure 6.1, assume the largest hexagon that can be inscribed in it. The area of such a hexagon was calculated in Chapter 2, considering the spherical shape of the Earth. For the sake of simplicity, ignore the spherical effect and calculate the approximate area of the hexagon inscribed in the circular service area with the radius r_0 (Figure 6.2) as

$$A_{hex} \approx 6r_0^2 \sin 30° \cos 30°$$

$$= \frac{3\sqrt{3}}{2} r_0^2 \tag{6.4}$$

Assuming nonoverlapping service areas after application of the TPC method, the ratio of a service area after application of the method to the service area before application is

$$\frac{A_{TPC}}{A_{hex}} = \frac{\pi (r_0 \rho)^2}{3\sqrt{3}\, r_0^2 / 2} = \frac{2\pi \rho^2}{3\sqrt{3}} \tag{6.5}$$

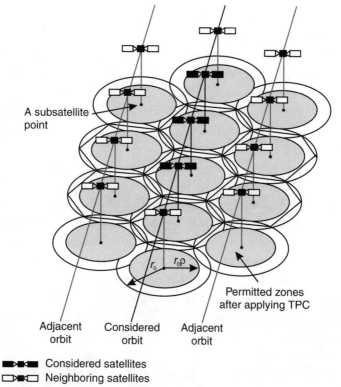

A subsatellite point

r_0 $r_0\rho$

Permitted zones after applying TPC

Adjacent orbit   Considered orbit   Adjacent orbit

■▶■◀■ Considered satellites
▭▶■◀▭ Neighboring satellites
r_0 : Radius of the service areas before applying the transmit permission control
ρ : Parameter of the transmit permission control scheme ($0 \leq \rho \leq 1$)

Figure 6.1   Configuration of the service areas of the satellites before and after application of the TPC scheme.

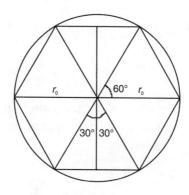

r_0 60° r_0

30° 30°

Figure 6.2   Calculating the area of a hexagon inscribed in a circle.

where A_{TPC} denotes the service area of a single satellite after employment of the TPC method. For example, if we employ the TPC with $\rho = 0.7$, the service areas are reduced to about 0.6 of those of the original areas. Because of the spherical shape of the Earth, the reduction in each service area is less than the one shown in (6.5), which can be calculated according to the mathematics given in Section 2.1.

It should be clear that selecting small values of ρ means most users have no permission most of the time; on the other hand, values of ρ close to 1 make the system have the characteristics almost the same as those of the system without such a control. Therefore, there should be an optimum value of ρ that provides the largest improvement in system performance. The next subsection discusses the performance of the system employing the TPC scheme and the selection of ρ; after that, we show the existence of such an optimum value.

6.1.3 Throughput performance of transmit permission control

Subsection 6.1.2 explained the protocol of the TPC scheme and briefly presented expectations on improvement in throughput of the system after application of this scheme. To analyze the performance of the scheme in detail, we use the measures of a single-satellite throughput, the total throughput, and the normalized total throughput similar to the ones explained in Chapter 4. To review those definitions, recall that the throughput of a satellite, say, the ith satellite, is the number of packets successfully received at that satellite and denoted by ξ_i; the summation of throughputs of three succeeding satellites is the total throughput; and the normalized total throughput for the three satellites is calculated by dividing the total throughput by the total number of users, N_u, as

$$\xi_{norm} = \frac{\xi_{i-1} + \xi_i + \xi_{i+1}}{N_u} \tag{6.6}$$

At each LEO satellite, the capture probability (i.e., the probability of packet success) of a given packet, namely, the tagged packet, depends not only on the power of multiple-access interference caused by simultaneous transmissions from the users in the same service area but also on the power of interference caused by transmissions of users in the service areas of adjacent satellites as well as the power of a constant background thermal noise. The reason is that in our system all users use the same frequency spectrum, and the signal of a user can arrive at a satellite and add to the level of multiple-access interference if that user is in the line of sight of that satellite. The only difference between the two types of interference is their power levels: Packets from users in the same service area as the tagged packet reach the satellite at equal levels, while packets from users in neighboring service areas arrive with power levels that depend on their distances and relative antenna gains.

With the explanation given in the Subsection 6.1.2, it may be clear that the only difference between the throughput analysis given in Chapter 4 and those that will be used in the TPC scheme is in the probability of packet transmissions in each time slot. That is, for the former analyses, in each slot every user transmits a packet with the probability of q. In the latter case we have the new assumption that every user whose propagation loss to the nearest satellite is less than l_{max} may send a packet in a slot with the probability of q.

Let us examine, via numerical examples, the throughput performance of the proposed scheme in both uniform and nonuniform traffic situations. We will show first that in both situations of traffic distribution the scheme can improve the peak of the total throughput. After that, by focusing on a nonuniform traffic situation that realizes a heavy traffic situation for one of satellites, we will show the ability of the method to improve the throughput of that satellite and hence the total throughput performance. Finally, we will discuss the selection of the maximum propagation loss, l_{max}, or its consequent parameter, ρ. Throughout these examples, a typical circular-orbit LEO satellite system with 11 satellites in each orbit flying at the altitude of 800 km is assumed. For analytical limitations, N_u is assumed to be 100.

To show the effects of the proposed scheme on the performance of the system, let us first assume an arbitrary value for ρ. This arbitrary selection makes no sense at the moment; however, it can exhibit some features of the TPC scheme. For two cases of relatively low multiple-access capabilities of $K(\mu_c) = 30$ and $K(\mu_c) = 10$, Figures 6.3 and 6.4, respectively, show the normalized total throughput characteristics at uniform and nonuniform traffic distributions when the TPC scheme with $\rho = 0.7$ is employed. The two figures correspond to the characteristics of

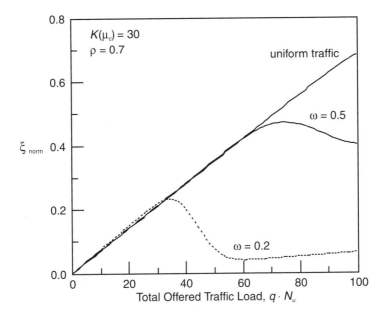

Figure 6.3 Throughput performance with the TPC scheme ($\rho = 0.7$) and a medium value of multiple-access capability.

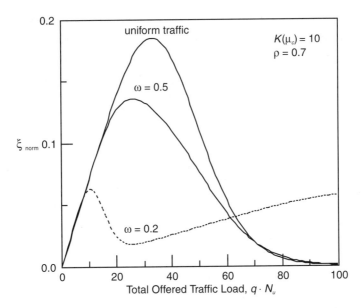

Figure 6.4 Throughput performance with the TPC scheme ($\rho = 0.7$) and a small value of multiple-access capability.

the conventional spread-slotted Aloha shown in Figures 4.11 and 4.12 after application of the TPC scheme. Comparing Figures 6.3 and 6.4 to the figures in Chapter 4, we can say that although each service area is reduced to about 60% of its original, applying the proposed scheme exhibits higher peak throughputs in both uniform and nonuniform traffic cases. In either case, the scheme shows improvement not only in enhancing the peak of the throughput curves but also in expanding the curves on the offered traffic load axis. Although we do not present any analysis on the stability characteristics of the system, by similarity of the multiple-access scheme with conventional slotted Aloha, we can expect that the expansion of the throughput curves on the offered traffic load axis, achieved by the proposed method, makes the system exhibit improved stability as well [12].

The improvements achieved by the scheme are due to different reasons. The prohibition from transmission of a portion of the users in the service area of each satellite is the main factor, since the power of interference of those users is the same as the power of the signal, S. The omission of interference from a portion of users of adjacent satellites is the second reason. The powers of interference of the latter users depend on the power loss factor β and their distances to the satellites.

As mentioned, the selection of ρ in the previous examples was arbitrary. An optimum selection of ρ, however, can be made by a trade-off between the level of traffic load and the degree of performance improvement. A decrease in the value of ρ results in a decrease in the number of users permitted to transmit in each service area. That means the number of simultaneous transmitting packets decreases, which is not necessary in light traffic loads. On the other hand, in heavy traffic loads restricting some portion of users from transmission improves the probability of packet success and thus the total throughput. Since in calculation of the total normalized throughput the total number of users, including both permitted and nonpermitted users, is considered, it is possible to find a proper value of ρ.

To show the effect of the selection of ρ on the performance of the system, let us focus the evaluation on a special case of $K(\mu_c) = 30$ and $\omega = 0.5$, which is a heavy traffic situation and hence a low-capture probability for one of the satellites.

Figure 6.5 shows the effect of changing the parameter of the TPC method ρ on the normalized total throughput of the system. As shown in the figure, decreasing the value of ρ from 1 (i.e., no transmit denials) to about 0.5 shows improvement in the maximum value of throughput curves. For traffic loads less than about 50, however, the system without the proposed method shows better characteristics. The reason is that in light traffic loads all the transmissions can be serviced with high probability of success, and prohibiting some users from transmission does not improve the probability of packet success of the others and only decreases the total number of the packets on air and hence the total throughput. On the other hand, for traffic loads higher than 50, the probability of packet success is relatively low, and this prohibition increases the probability of packet success for the permitted users, so that larger throughput is achievable. Small values of ρ can give improvement only in heavy offered traffic loads.

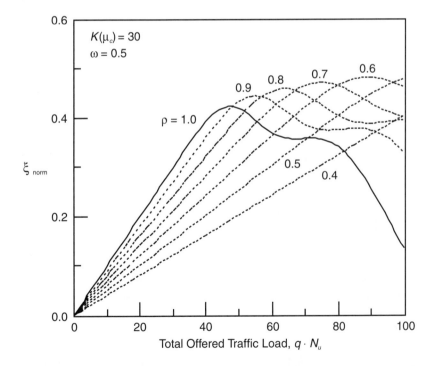

Figure 6.5 Effect of selection of ρ on normalized total throughput performance.

To exhibit the relation between the selection of ρ and the offered traffic load more clearly, we have examined the changes in normalized total throughput at different offered traffic loads when the parameter of the TPC scheme, ρ, is changed. Figure 6.6 compares the normalized total throughput for different values of offered traffic load $q \cdot N_u$ as a function of ρ. The point illustrated by Figure 6.6 is that for each traffic load there is an optimum value for ρ that results in maximum improvement of the throughput performance. That means there is a close relation between the selection of ρ and the offered traffic load region in which the system works: A system working at heavy traffic load requires smaller values of ρ, while for a system in which a light offered traffic region is designed, the values of ρ near 1 is enough. The figure also illustrates that using small

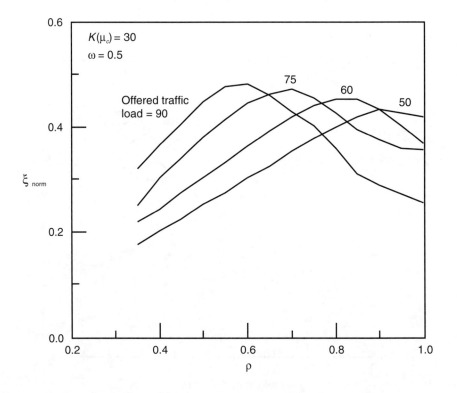

Figure 6.6   Effect of selection of ρ on normalized total throughput performance in different offered traffic loads.

values of ρ at high offered traffic situations significantly improves the performance of the system.

As mentioned before, in LEO satellite systems, the satellites are in a continuous motion. Thus, it is necessary to consider dynamic performance of the system while it is experiencing high traffic load areas in the satellites' paths. To do that, let us evaluate the dynamic performance of the system by a new measure, namely, the average normalized total throughput. The average normalized total throughput is defined as the average value of the normalized total throughput of the three succeeding satellites covering a particular area on the Earth during a specific period of time. If we denote, as before, the total number of the satellites in one orbit by N_s, the averaging procedure is done in the $2\pi / N_s$ fraction of the orbit in which the density function of the users is centered. During that period, the satellites experience different traffic-load situations. The case of one satellite with high traffic load and the two others with light traffic loads described until now is one of these situations.

Figure 6.7 shows the average normalized total throughput with the same parameters as in Figure 6.5, as a function of total offered traffic load and for different values of ρ. Comparing the two figures, we can see almost the same behavior in improvement of the peak value of the throughput by employing the TPC scheme as the previously considered instantaneous throughputs. However, interesting differences at high offered traffic loads between the curves of the two figures also are recognizable. That is, at high offered traffic loads, the values of the average total throughput are much higher than the instantaneous values in Figure 6.5. The reason is that in Figure 6.5 a heavily populated area is serviced by a single satellite, that is, only the worst case is considered. However, after averaging the instantaneous throughputs, other situations, for example, the case when the peak of the traffic is between two satellites, also are considered.

6.1.4   Average delay performance of transmit permission control

Another important measure of performance in a packet communications system is the average delay performance, which shows on average the

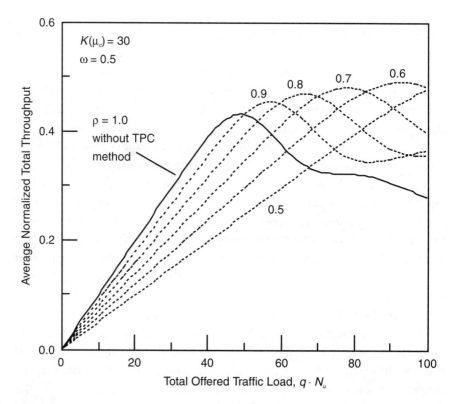

Figure 6.7   Effect of the TPC method on the average normalized total throughput.

required time for successful delivery of a packet in the system. If the spread-slotted Aloha scheme is employed in a communication system with a negligible value of propagation delay, such as in microcellular systems or in a LAN, the average delay is due mainly to the number of retransmissions of collided packets because of a high level of interference experienced by those packets. After a packet is involved in a collision, the sender of that packet should retransmit the same packet repeatedly after a randomly selected period until an acknowledgment eventually is received. The random retransmission delay hopefully avoids the collision of the same packets in future retransmissions. In a low-throughput situation, the number of such collisions and, hence, the number of retransmissions increase, and then average delay becomes longer.

In LEO satellite systems, although the propagation delay between users and satellites is much smaller than that in a geostationary satellite system, if the number of retransmitted packets increases, the total propagation delay becomes on the order of the packet length and thus is not negligible. If we improve the probability of packet success, or the throughput, the number of collisions decreases, and we can expect an improved average delay performance of the system as well. The proposed TPC scheme, as shown before, improves the throughput characteristics; therefore, it may improve the delay performance, too. However, because application of this method may cause another delay time, that is, a delay due to the waiting time to get the permission for transmission for a portion of users, we should consider the average delay performance in the evaluation of the method in addition to the throughput performance. A trade-off between the average delay and the throughput performance may determine the optimum degree of the TPC scheme.

Average delay is defined, in general, as the average time elapsed from the moment a packet is generated by a user to the moment the entire packet is received successfully at a satellite [13–16]. Here, we consider two kinds of average delay. The first one is the average delay of the packets generated by the users in the service area of each satellite, say, the ith satellite, and which is denoted by $\overline{\Delta}_i$. Because the expected number of users and the throughput in the service areas of satellites in nonuniform traffic distribution are different, the average delay in different service areas also is different. Therefore, we define the normalized average delay, $\overline{\Delta}_{norm}$, as the average delay of the packets generated in any service area. For example, considering three satellites, we have

$$\overline{\Delta}_{norm} = \frac{\overline{\Delta}_{i-1} \, E\{N_{i-1}\} + \overline{\Delta}_i \, E\{N_i\} + \overline{\Delta}_{i+1} \, E\{N_{i+1}\}}{N_u} \quad (6.7)$$

where $E\{N_i\}$ is the expected number of users in the service area of the ith satellite. Obviously, in uniform traffic distribution, $\overline{\Delta}_{norm}$ will be equal to the average delay in each of the service areas.

In this subsection, we evaluate the average delay performance of the TPC scheme. That analysis also can be used in the case of a conventional spread-slotted Aloha scheme with the assumption of an infinity value for l_{max}. In Subsection 6.1.3, it was assumed that any user with a propagation loss less than l_{max} to the nearest satellite transmits a packet in each time slot with the probability q. That assumption leads to a binomial composite arrival distribution, including both the new originated packets and the retransmitted ones. In other words, it is assumed that the new originated packets and the retransmitted packets have the same packet generation statistics. Although in general, for practical implementation, the probability of transmitting a new originated packet, p_o, is smaller than that of a retransmitted packet, p_r, it has been shown that the assumption of $p_o = p_r = q$ gives major simplifications and exact-enough results [14]. The need for retransmission is due to the loss of the packet because of an excessive interference level.

Any user wishing to transmit a packet first should check the permission for transmission according to the protocol of the TPC scheme. A user in a nonpermitted zone should wait until a satellite comes close enough so that the propagation loss to that satellite becomes less than l_{max}. The probability of being a user in nonpermitted zones of each service area is related to the percentage of the permitted zones in that service area and also to the probability density function of the location of users and for the ith satellite, denoted by p_{np_i}. Note that without employment of the TPC method, $p_{np_i} = 0$ in all service areas. The waiting time for obtaining permission is denoted by τ_{wp} and can have any value between 0 and $2r_0(1 - \rho)/v$, where r_0 is the radius of the service areas and v is the ground speed of the satellites; when ρ is enough large, it is assumed that the average value of τ_{wp} is $\overline{\tau}_{wp} = r_0(1 - \rho)/v$. Therefore, the average waiting time related to the TPC in the service area of the ith satellite is $\overline{\tau}_{wp}p_{np_i}$.

A user waits for an acknowledgment from the destination satellite before clearing the packet. Because a slotted case is considered, the packet can be generated at any point during a slot, yet the user has to wait until the beginning of the next slot before attempting a transmission. The time between packet generation and the start of the next slot is represented by τ_{pg} and can have any value between 0 and τ_p, the packet duration, with equal probability. Hence, the average time from the generation of the packet until the entire packet enters the channel equals $\overline{\tau}_{wp}p_{np_i} + \tau_p/2 + \tau_p = \overline{\tau}_{wp}p_{np_i} + 3/2\tau_p$. For the purpose of comparing the average delay performance in the conventional DS/SSA system with a system that employs the TPC scheme and for the sake of simplicity in calculations, zero-guard time is assumed here.

The packet travels through the uplink satellite channel and is subjected to the one-hop satellite delay, T_d. Although T_d differs according to the location of the transmitting user in a service area, for a satellite system its variance is small so that an average value of T_d is used in numerical examples. In the case of successful reception of the packet by the satellite, on average $\tau_d = \bar{\tau}_{wp}P_{np_i} + 3/2\tau_p + T_d$ seconds elapse between generation of the packet and its full acceptance by the satellite. In this case, the satellite sends an acknowledgment packet with a duration of τ_{ack}, and the user receives that packet after $T_d + \tau_{ack} + \tau_{process}$ seconds, where $\tau_{process}$ is the required satellite processing time. Here we assume that τ_{ack} and $\tau_{process}$ are negligible compared to other time durations. From the moment of sending a packet, the user waits for another T_d seconds, expecting to receive the acknowledgment packet. If the attempt is successful, the user clears the packet. If an acknowledgment packet is not received, the user considers the packet lost and starts the process of reattempting transmission. It is reasonable to assume that the probability of loss of the acknowledgment packet is very small. If the packet is lost, the user waits τ_w seconds, a random retransmission delay time with an average of $\bar{\tau}_w$. The retransmission procedure is repeated until the user receives an acknowledgment. In any steps of the retransmission, the user may find the unpermitted condition and then should wait until permission is obtained.

For the scenario described here, the average delay of the packet in the service area of the ith satellite can be expressed as

$$\bar{\Delta}_i = \tau_d \, p_{suc_i} + (\tau_d + \bar{\tau}_{tw})(1 - p_{suc_i})p_{suc_i} + (\tau_d + 2\bar{\tau}_{tw})(1 - p_{suc_i})^2 p_{suc_i} + \cdots$$

$$= \sum_{k=0}^{\infty} (\tau_d + k\,\bar{\tau}_{tw})(1 - p_{suc_i})^k \, p_{suc_i} \tag{6.8}$$

where $\bar{\tau}_{tw}$ is the average time elapsed between the end of the first transmitted packet and the end of the first retransmitted packet or between the ends of any two successive retransmitted packets, at the satellite, equal to $\bar{\tau}_{tw} = T_d + \bar{\tau}_w + \tau_d$, and P_{suc_i} is the average of the probability of packet success in the service area of the ith satellite. Utilizing the properties of an arithmetic-geometric series, (6.8) can be simplified into

$$\bar{\Delta}_i = \tau_d + \bar{\tau}_{tw} \left(\frac{1}{P_{suc_i}} - 1 \right) \tag{6.9}$$

Let us now examine the effect of the TPC scheme on the average delay performance of the system. The following examples assume $\tau_p = 0.1\,\text{s}$ and $\bar{\tau}_w = 3\tau_p$. Figure 6.8 shows the normalized average delay, $\bar{\Delta}_{norm}$, versus

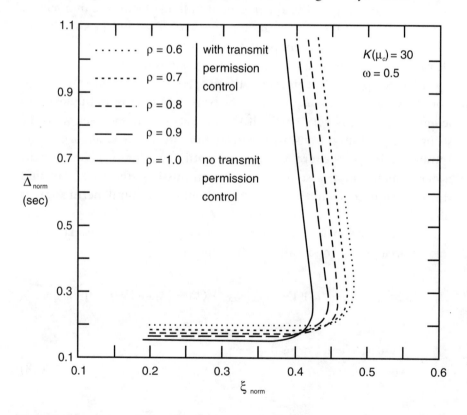

Figure 6.8   Comparison of the average delay-throughput performance of the systems with and without the TPC scheme.

the normalized total throughput under the same conditions as in Figure 6.5, again with ρ as a parameter. At initial low throughput regions, which correspond to light offered traffic loads, employing TPC seems to enlarge the average delay, similar to the discussion of the throughput performance in Figure 6.5, because of the prohibition from transmission of a fraction of the users that can be serviced successfully. At higher throughput, however, the average delay performance of the system employing the TPC becomes superior to that of the system without this scheme. In the higher throughput, employing the proposed scheme decreases the large number of retransmissions and hence shortens the average delay. Figure 6.8 suggests that an adaptive selection of ρ can offer improved average delay, similar to the suggestion given for the throughput performance according to Figure 6.5, in a wide range of change in offered traffic load. Note that employing the TPC with a fixed value of $\rho = 0.7$ enlarges the average delay at most on the order of half a packet length. The half-packet-length delay is an expected value even in conventional (unspread) slotted Aloha systems.

Having analyzed the average delay, it now is simple for us to consider the effect of nonuniformity in traffic distribution on the average delay performance of the LEO satellite systems. The average delay performance can show the stability of the packet communication system easily. Figure 6.9 shows the average delay-throughput performance of a LEO satellite system in different traffic distributions. In the figure, a conventional spread-slotted Aloha system with a relatively low value of multiple-access capability is considered.

Comparing the performance in a uniform traffic situation with that in a nonuniform traffic distribution, Figure 6.9 illustrates that in a uniform traffic situation the system exhibits much more stability than in nonuniform traffic situations. That is, in a uniform traffic situation when the level of offered traffic load to the system increases, the relative increase in average delay time occurs much more smoothly than in the case of nonuniform traffic distributions. That is in addition to the higher total throughput that the system can handle in uniform traffic conditions compared to nonuniform situations. The reason for having a more stable system in uniform traffic is that at nonuniform traffic situations the expected number of users in the service area of one of the satellites is very

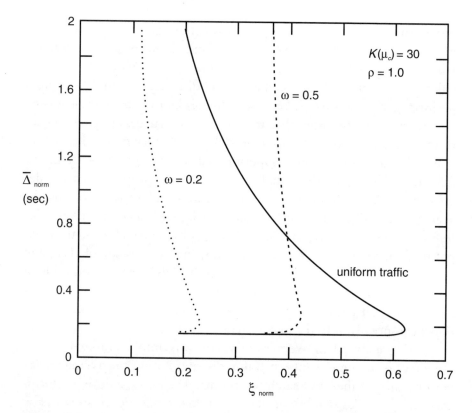

Figure 6.9 Average delay-throughput performance of the system without the TPC at different traffic situations.

large; thus, the number of retransmitted packets is very large. Therefore, the increase in the average delay after collisions in these traffic situations becomes sharper compared to that in the uniform traffic situation.

The new results derived from Figure 6.9 and the results on the throughput performance in different traffic distributions given previously show that traffic nonuniformity affects the performance of the system significantly. By examining either the throughput or the average delay performance, we conclude that traffic nonuniformity is a serious problem for future global LEO satellite systems and should be considered during the design of any LEO satellite system for global communications services.

6.2 Transmit permission control scheme: Fading channel

Until now, a nonfading or ideal land-mobile satellite channel was assumed. That assumption is a proper simplification, especially for satellite systems with relatively high elevation angles. In the case of LEO satellite systems, however, low elevation angles on the order of 10 degrees often are considered. In this section, we consider the fading problem in our LEO satellite system and evaluate the differences in performance of the system compared with considered nonfading channel assumption. Moreover, we analyze the performance of the TPC scheme in a channel that suffers from variations of the received signal power due to the fading.

We will start the discussion of the fading problem in satellite channels with an overview of the factors that affect the power of the signals in satellite systems [17–25]. Satellite communications with land-mobile terminals suffer from strong variations of the received signal power due to signal *shadowing* and *multipath fading*. Shadowing of the satellite signal by obstacles in the propagation path, such as buildings, bridges, and trees, results in attenuation over the total signal bandwidth. The attenuation increases with carrier frequency; that is, it is more marked at *L*-band than at *ultra high frequency* (UHF). For low satellite elevation, the shadowed areas are larger than for higher elevations. Multipath fading occurs because the satellite signal is received not only via the direct path but also after being reflected from objects in the surroundings. Due to their different propagation distances, multipath signals can add destructively, resulting in a deep fade.

Subsection 6.2.1 models those two kinds of variations in signal powers and shows how the variations affect the results obtained up to this point. Because TPC schemes prohibit transmissions from users with low elevation angles, we will show that the TPC scheme can solve the problem of fading in LEO satellite systems to some degree.

6.2.1 Fading channel model and analysis

A number of models have been proposed for satellite mobile fading channels. Loo has proposed a model suitable for rural areas, which assumes that the received signal consists of a shadowed direct component and a scattered diffuse component [19]. The direct component is affected by log-normal shadowing, while the diffuse scattered component undergoes Rayleigh nonselective fading. Lutz et al. introduced a two-state model that is Rice in open areas and Rayleigh-log-normal otherwise [21]. Vucetic and Du presented a Markov M-state channel model, where a Markov chain is used to model long-term state transitions [25]. Also, Vucetic et al. presented a Markov channel model with M-states for *intermediate circular orbit* (ICO) satellite systems, where each state is a linear combination of log-normally distributed direct and ground-reflected signals and a Rayleigh-distributed scattered signal [18].

Although all these models, as well as many others, have some specifications on the satellite system, here we use the model proposed by Lutz et al., which includes statistical results for different satellite elevations and different environments and seems to be useful in LEO satellite systems [21]. This model is described by the time-share of shadowing B; the fraction of time that the signal is shadowed. During the shadowed interval, the channel is modeled as log-normal frequency nonselective Rayleigh fading. In the unshadowed period of time-share $1 - B$, the channel is modeled as frequency nonselective Rician. The resulting probability density function of the received signal power normalized to the power of the signal in the absence of either fading or shadowing is then given by the mixture density

$$f_V(V) = 2(1 - B) \sqrt{V}ce^{-c(V + 1)} I_0(2c\sqrt{V}) + 2B\sqrt{V}ce^{-cV} \quad (6.10)$$

where c is the direct-to-multipath signal power ratio (Rice-factor) and $I_0(\cdot)$ is the modified Bessel function of the first kind and zeroth order. The parameter V can be thought as the factor that shows the effects of fading: In a nonfading channel, it is a unity constant value; in a fading channel, it is a random variable with the probability density function given in (6.10). Note that the expected value of the normalized received power in the absence of either fading or shadowing is unity.

Although this channel model can be used for both directions, uplink and downlink, to make the effect of fading on the uplink clear, here we assume nonfading downlink channels. That way, users will have the exact information on their distances to the satellites and transmit with the proper power level given in (6.3). Lutz et al. offer a complete discussion of the parameters c and B for different satellite elevations and different environments [21]. This section considers $c = 10$ as a typical value for LEO satellite channels and $B = 0.3$ as a compromised value between large values (for an urban area) and small values (for a suburban or rural area), as used in [22,23].

According to (6.3), a user, namely the jth one, in the service area of the ith satellite and in the distance $d_{i,j}$ from it, transmits a packet by the power $T_{i,j}$ to ensure the received power level S at that satellite in the absence of fading. However, because of the fading, the level of the received signal from that user at the ith satellite may be different from the level S according to the level of fading. In this case, the signal of the jth user reaches the ith satellite with the level

$$S_{i,j} = T_{i,j}\, \kappa d_{i,j}^{-\beta}\, V_{i,j} \tag{6.11}$$

where $V_{i,j}$ is a random variable with the probability density function given in (6.10). The signals of the users in the service areas of the neighboring satellites in the line of sight of the ith satellite arrive at the ith satellite with power levels related to the distances of each user from the two satellites and also the level of fading. Without lack of generality, assume that the kth user in the service area of the $(i+1)$st satellite is one of those users and has the distance $d_{i,k}$ to the ith satellite and the distance $d_{i+1,k}$ to the $(i+1)$st satellite. The user transmits packets by the power $T_{i+1,k}$ to ensure the received power level S at the $(i+1)$st satellite in the absence of fading. Because of fading channel, the signal reaches the ith satellite with the power

$$S_{i,k} = T_{i+1,k}\, \kappa d_{i,j}^{-\beta}\, V_{i,k} \tag{6.12}$$

where, again, $V_{i,k}$ is a random variable with the probability density function given in (6.10).

As mentioned in Chapter 4, to calculate the throughput, it is necessary to find the probability of packet success in the service area of each satellite. This probability was derived in Section 4.2.3 for nonfading satellite links. With the fading uplink channels defined in (6.10), now we should find the probability of packet success, $P_{C,i}$ for the fading environment. In the nonfading environment discussed in Chapter 4, the power level of the target packet is the fixed value S, which appears in (4.18), (4.21), and (4.37). On the contrary, in the fading environment, it is a random variable according to the random variable $V_{i,j}$, as described by (6.11). On the other hand, the total power of the interfering packets, I_i, is a random variable even without fading, as are the locations and thus the powers at the satellite of $m - n$ users of the neighbor satellites. When fading is considered, the total power of the $m - n$ users are still random variables, but not only by the locations of the users but also because of fading, as is denoted in (6.12). In addition to the $m - n$ users outside the service area, the level of the interference from each of the remaining interfering users inside the service area also is not constant, but random because of the fading, as in (6.11). Considering those differences in the fading environment, we derive numerically the packet success probability, $P_{C,i}$, and find the throughput characteristics with (4.14). Note that in the calculation, with the assumption of performing the analysis during a short period of time and a reasonably high bit rate, we assume that the fading varies slowly compared to the bit rate; hence, the received signal power is considered constant during one symbol interval.

6.2.2 Numerical examples of the performance of the system

With the mathematics given in Subsection 6.2.1, we now can evaluate the performance of our LEO satellite system in a fading channel. For the same parameters as in Figure 6.5, Figure 6.10 shows the normalized total throughput before and after application of the TPC scheme now in the fading channel with $c = 10$ and $B = 0.3$. For the TPC case, the two values of $\rho = 0.7$ and $\rho = 0.6$ are shown.

First we compare the performance of the conventional systems (i.e., without the TPC scheme) in fading and nonfading satellite channels. Comparing the results of the case $\rho = 1$ shown in Figures 6.5 and 6.10, we can observe some throughput enhancements at heavy traffic loads in the fading channel. One reason is that at heavy traffic loads the effect of fading acts positively to improve the probability of the packet success by

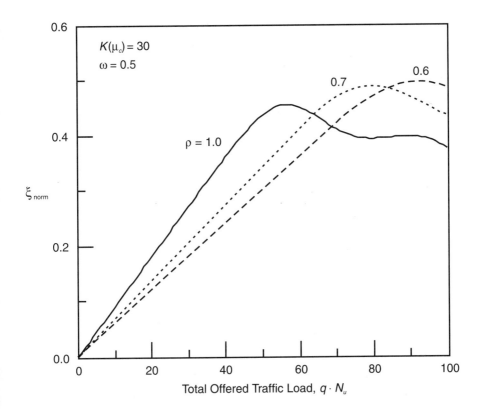

Figure 6.10 Throughput performance with and without the TPC scheme in fading channel.

decreasing the power of interference and thus to improve the throughput. On the other hand, at light traffic loads, fading decreases the total throughput slightly. At light offered traffic loads, the effect of fading on the desired signal dominates over the level of interference, because in this case the number of interferers is small.

For the case utilizing the TPC scheme, as seen in Figure 6.10, in the fading channel the performance of the system at heavy traffic loads still can be improved. However, after application of the TPC scheme, because the number of transmitting users decreases, the achievable enhancement due to fading in the TPC case becomes less than in the case without TPC.

Moreover, with the TPC, the permitted users are mostly the ones with high elevation angles for which the effect of shadowing on attenuation of the power of the received signals is less. According to the results shown in both fading and nonfading channels, we can conclude that assuming a fixed value of ρ at all offered traffic loads is not a good idea and that an adaptive selection of the maximum allowable propagation loss l_{max} may result in an optimum improvement at all traffic loads. Section 6.3 proposes two schemes to adjust the value of ρ in an adaptive manner and show their performance improvement in the case of LEO satellite systems.

6.3 Adaptive transmit permission control schemes

The results shown in Figures 6.5 through 6.8 and Figure 6.10 illustrate the dependency of the effectiveness of the TPC method on the level of the offered traffic load: Higher offered traffic loads need smaller values of ρ to have better performance, and lighter traffic loads require larger values of ρ. That fact implies that an adaptive selection of ρ according to the total offered traffic load can improve the throughput performance of the system. In addition, in the case of nonuniform traffic distribution, if we see the throughput characteristics of individual satellites, for example, the satellite over the dense traffic area and its neighbor satellites over the sparse traffic areas, we can expect to find better performance by adaptive selection of ρ for each satellite according to its level of the offered traffic load. For example, the throughput performances of a satellite with heavy traffic load and a satellite with light traffic load in an assumed nonfading channel are shown in Figures 6.11 and 6.12, respectively.

According to Figure 6.12, it is clear that for a satellite with light traffic loads and a given value of multiple-access capability it is not a good idea to select small values of ρ, such as $\rho < 0.8$, because in light traffic situations all the transmissions can be received successfully at the satellite with high probability of success, and prohibition from transmission to such a satellite only reduces the number of packets on air without any considerable improvement in packet success probability, thus degrading the

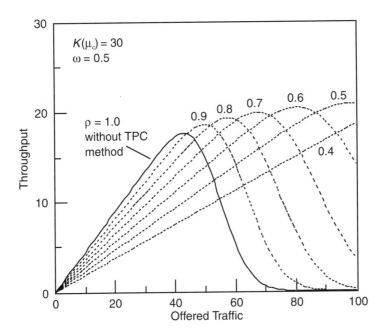

Figure 6.11 Effect of the TPC on the throughput performance of the satellite over the dense traffic area.

total throughput of that satellite. On the other hand, according to Figure 6.11, for a satellite with heavy traffic loads, it is better to select smaller values of ρ, which means limiting its service area more. For such a satellite, selecting even very small values, such as $\rho < 0.6$, still improves the peak of the throughput performance.

The idea given from these two typical examples, that is, the selection of ρ according to the traffic loads of individual satellites, in addition to the idea in Figure 6.6, that is, the selection of ρ according to the total offered traffic load, induce the two possible adaptive methods on the selection of ρ. We refer to those methods as *adaptive* TPC (ATPC) methods [26,27].

6.3.1 ATPC method 1

The first method considers the selection of ρ according to the change of the total offered traffic load of the three satellites. Similar to the basic

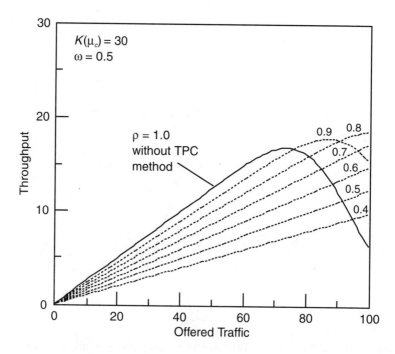

Figure 6.12 Effect of the TPC on the throughput performance of the satellite over the sparse traffic area.

TPC method, the reduction in the service areas of all satellites is the same, by the factor ρ. However, different from the basic method, the value of the ρ is not constant but changes according to the change of the total offered traffic load. In this method, according to the total offered traffic load statistics, which are estimated from the statistics of the previous time slots, the optimum value for ρ is calculated so the maximum total throughput can be achieved and the users are informed via downlink information channels. This value of ρ is common in all service areas and is updated regularly, depending on how often the level of traffic load changes. The result is the equal reduction or enhancement of the service areas compared to the ones established in the last time slots. In this scheme, regardless of the different traffic loads offered to the satellites, the sizes of the service areas of all satellites are kept the same even after the method is applied.

6.3.2 ATPC method 2

The second method considers the selection of ρ according to the change in the offered traffic load of individual satellites. Similar to the first method, the statistics of the traffic load are used for determining the optimum value for ρ. However, different from the first method, the decision is not common for all satellites and is performed by each one and is valid only for that satellite. The optimum value for ρ in this method is the value that makes the throughput of each satellite maximum. That makes the satellite with lighter traffic loads select larger values for ρ. This method is especially effective for the case of nonuniform traffic situations, in which different satellites have different total traffic loads. Obviously, in a uniform traffic scenario, the performance of this method agrees with the performance of the first method, and the realization technical parameters select the best method. After this method is applied, the service area of the satellite with the higher traffic load becomes smaller than that of the satellite with the lighter traffic load.

6.3.3 Performance of ATPC methods

Let us compare the performance of the system without the TPC method with the ones employing ATPC methods 1 and 2. Figure 6.13 shows the performance of the system under the same conditions as Figure 6.5. Employing the first adaptive method maintains good performance at light offered traffic load by disabling the TPC (i.e., selecting ρ = 1) and improves the throughput at higher offered traffic loads by gradually decreasing the value of the ρ. As the offered traffic load increases, the value of ρ decreases equally for all three satellites. However, as shown in Figures 6.11 and 6.12, a decrease in ρ according to the total traffic load is proper mostly for the satellite with high traffic load; a large decrease in the value of ρ is not suitable for the light traffic satellites. Contrarily, by employment of the second method, the value of ρ can be determined for each satellite independently; hence, the performance of the system by the measure of the normalized total throughput improves more compared with the first adaptive method. In the latter method we improve the throughput of each satellite separately; hence, the method exhibits better total throughput at the whole range of the offered traffic

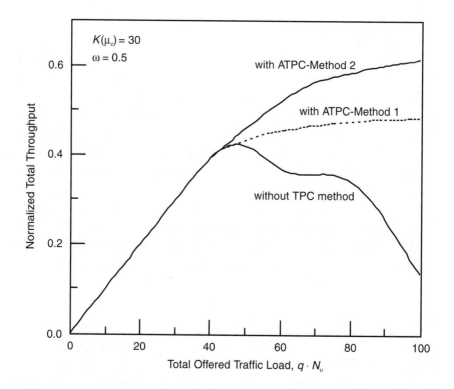

Figure 6.13 Comparison of the performance of the systems without the TPC method, with ATPC-Method 1 and with ATPC-Method 2.

load. In other words, in the ATPC method 2, we assign a traffic load to each satellite near to the traffic level that can be serviced by that satellite. This is the reason for its better performance.

Although the improved characteristics of the second adaptive method are much more attractive than the first method, the implementation of the second method requires much more complexity. In the first method, the decision on the optimum value of ρ is based on the offered traffic load in the area serviced by a group of satellites, for example, in the model of three satellites, which changes slowly; hence, the change in ρ also should perform slowly. However, for the second method, especially in a nonuniform traffic situation, the offered traffic load to each satellite changes rapidly; then ρ must be changed often. A trade-off between such complexities and the differences in their performance improvement may determine the best method to be utilized in a real system.

6.4 Summary

For the LEO mobile satellite systems in which the relative location of each user to the satellites changes rapidly, in this chapter we proposed a TPC scheme in which at given intervals of time the permission for transmission is prohibited from a fraction of users. In this method, the permission condition of transmission for each user is determined according to its propagation loss to its observable satellite(s). The TPC scheme is applicable in direct-sequence spread-slotted Aloha multiple-access systems, and it can be considered as a means of reducing the power of multiple-access interference and hence improving the throughput performance of the system. It was shown that with a proper selection of the maximum propagation loss, beyond which a user is not allowed to transmit, the method can improve significantly the throughput performance of the LEO satellite communication system, in both uniform and nonuniform traffic situations. Moreover, the method is suitable for systems in which the satellites are faced with heavy traffic loads.

By calculating the average delay throughput performance in addition to the throughput performance, it was shown that although the method seems to enlarge the delay time for transmission of a given packet, because of the high ground speed of the satellites in a typical low-Earth orbit, the increase in the average delay time is only on the order of half of a packet length, similar to conventional slotted Aloha systems.

By introducing a fading satellite channel model, which considers both shadowing and multipath fading, and modifying the mathematics of Chapter 4, it was shown that at heavy traffic loads the fading decreases the power of interference received at each satellite and thus enhances the throughput characteristics. Moreover, it was shown that the improvement in the throughput performance of the TPC scheme is kept even under assumptions of a fading channel. Because in the TPC scheme users with relatively low elevation angles are prohibited from transmission, the method is interesting in fading channels.

In addition, this chapter proposed two new adaptive TPC schemes for direct-sequence spread-slotted Aloha multiple-access systems. The methods were applied on the uplinks of a LEO satellite communications system, again with the assumption of a nonuniform traffic distribution. It was shown that the methods significantly improve the throughput characteristics of the system compared to the conventional spread-slotted Aloha method and nonadaptive form of the TPC scheme and that the improved characteristics can be maintained in a wide range of the offered traffic load. The proposed methods are proper mostly for packet data communications, in which the discontinuity in the communication can be acceptable. If the number of satellites increases, the discontinuity periods become shorter and the method becomes applicable even in voice systems.

Although the mobility feature of LEO satellites makes the method practical here, the method can be applied in, for example, terrestrial cellular mobile communication systems, in which the base stations are fixed but the mobiles move around.

References

[1] Makrakis, D., and K. M. Sundaru Murthy, "Spread Slotted ALOHA Techniques for Mobile and Personal Satellite Communication Systems," *IEEE J. Select. Areas Commun.*, Vol. 10, No. 6, 1992, pp. 985–1002.

[2] Jamalipour, A., et al., "Transmit Permission Control Scheme for Spread-Slotted Aloha in LEO Satellite Communication Systems," *Proc. 2nd Asia-Pacific Conf. Commun. (APCC '95)*, Osaka, Japan, 1995, pp. 848–852.

[3] Jamalipour, A., et al., "Adaptive Transmit Permission Control Scheme for Spread-Slotted Aloha Mobile Satellite Communication Systems," *Proc. 18th Symp. Inform. Theory Appl. (SITA '95)*, Hanamaki, Japan, 1996, pp. 777–780.

[4] Gilhousen, K. S., et al., "Increased Capacity Using CDMA for Mobile Satellite Communication," *IEEE J. Select. Areas Commun.*, Vol. 8, No. 4, 1990, pp. 503–514.

[5] Viterbi, A. J., "A Perspective on the Evaluation Of Multiple Access Satellite Communication," *IEEE J. Select. Areas Commun.*, Vol. 10, No. 6, 1992, pp. 980–983.

[6] Kohno, R., R. Median, and L. B. Milstein, "Spread Spectrum Access Methods for Wireless Communications," *IEEE Commun. Mag.*, Vol. 33, No. 1, 1995, pp. 58–67.

[7] Jamalipour, A., et al., "Throughput Analysis of Spread-Slotted Aloha in LEO Satellite Communication Systems With Nonuniform Traffic Distribution," *IEICE Trans. Commun.*, Vol. E78-B, No. 12, 1995, pp. 1657–1665.

[8] Viterbi, A. J., "The Evaluation of Digital Wireless Technology From Space Exploration to Personal Communication Services," *IEEE Trans. Vehic. Technol.*, Vol. 43, No. 3, 1994, pp. 638–644.

[9] Jamalipour, A., et al., "Transmit Permission Control on Spread Aloha Packets in LEO Satellite Systems," *IEEE J. Select. Areas Commun.*, Vol. 14, No. 9, 1996, pp. 1748–1757.

[10] Gilhousen, K. S., et al., "On the Capacity of a Cellular CDMA System," *IEEE Trans. Vehic. Technol.*, Vol. 40, No. 2, 1991, pp. 303–312.

[11] Simpson, F., and J. Holtzman, "CDMA Power Control, Interleaving, and Coding," *Proc. 41th IEEE Vehic. Technol. Conf.*, St. Louis, 1991, pp. 362–367.

[12] Joseph, K., and D. Raychaudhuri, "Stability Analysis of Asynchronous Random Access CDMA Systems," *Proc. IEEE GLOBECOM '86*, 1996, pp. 1740–1746.

[13] Joseph, K., and D. Raychaudhuri, "Throughput of Unslotted Direct-Sequence Spread Spectrum Multiple-Access Channels With Block FEC Coding," *IEEE Trans. Commun.*, Vol. 41, No. 9, 1993, pp. 1373–1378.

[14] Raychaudhuri, D., "Performance Analysis of Random Access Packet-Switched Code Division Multiple Access Systems," *IEEE Trans. Commun.*, Vol. COM-29, No. 6, 1981, pp. 895–901.

[15] Ganesh, R., et al., "Performance of Cellular Packet CDMA in an Integrated Voice/Data Network," *Int. J. Wireless Infor. Networks*, Vol. 1, No. 3, 1994, pp. 199–222.

[16] Abdelmonem, A. H., and T. N. Saadawi, "Performance Analysis of Spread Spectrum Packet Radio Network With Channel Load Sensing," *IEEE J. Select. Areas Commun.*, Vol. 7, No. 1, 1989, pp. 161–166.

[17] Proakis, J., *Digital Communications, 2nd ed.*, New York: McGraw-Hill, 1989.

[18] Vucetic, B., L. Zhang, and I. Oppermann, "Modeling and Simulation of ICO Satellite Communication Channels," *Proc. Int. Symp. Inform. Theory, Appl. (ISITA '94)*, Sydney, Australia, 1994, pp. 493–497.

[19] Loo, C., "Measurements and Models of a Land Mobile Satellite Channel and Their Applications to MSK Signals," *IEEE Trans. Vehic. Technol.*, Vol. VT-35, No. 3, 1987, pp. 114–121.

[20] Vogel, W. J., and E. K. Smith, "Theory and Measurements of Propagation for Satellite to Land Mobile Communication at UHF," *Proc. 35th IEEE Vehic. Technol. Conf. (VTC '85)*, Boulder, CO, 1985, pp. 218–223.

[21] Lutz, E., et al., "The Land Mobile Satellite Communication Channel—Recording, Statistics and Channel Model," *IEEE Trans. Vehic. Technol.,* Vol. 40, No. 2, 1991, pp. 375–386.

[22] Monk, A. M., and L. B. Milstein, "Open-Loop Power Control Error in a Land Mobile Satellite System," *IEEE J. Select. Areas Commun.,* Vol. 13, No. 2, 1995, pp. 205–212.

[23] Vojcic, B. R., R. L. Pickholtz, and L. B. Milstein, "Performance of DS-CDMA with Imperfect Power Control Operating Over a Low Earth Orbiting Satellite Link," *IEEE J. Select. Areas Commun.,* Vol. 12, No. 4, 1994, pp. 560–567.

[24] Corazza, G. E., and F. Vatalaro, "A Statistical Model for Land Mobile Satellite Channels and Its Application to Nongeostationary Orbit Systems," *IEEE Trans. Vehic. Technol.,* Vol. 43, No. 3, 1994, pp. 738–742.

[25] Vojcic, B. R., and J. Du, "Channel Modeling and Simulation in Satellite Mobile Communication Systems," *IEEE J. Select. Areas Commun.,* Vol. 10, No. 8, 1992, pp. 1209–1218.

[26] Jamalipour, A., et al., "Adaptive Transmit Permission Control on Spread-Slotted Aloha Packets Applicable in LEOS Systems," *IEICE Trans. Commun.,* Vol. E79-B, No. 3, 1996, pp. 257–265.

[27] Jamalipour, A., et al., "Throughput Performance Improvement of Spread-Slotted Aloha in Mobile Satellite Communications," *Proc. IEEE Int. Conf. Commun. (ICC '96),* Dallas, TX, 1996, pp. 1740–1744.

7

Further Considerations in LEO Satellite Systems

HE FIRST SECTION of this chapter proposes another method for improving the throughput performance of a LEO satellite system employing spread-spectrum techniques. We start the analysis in Section 7.1 by discussing the effects of multiple-access interference on throughput performance of a direct-sequence spread-spectrum LEO satellite communications network. To recognize the effects of interferences when their sources are either inside or outside the service area of a satellite, we develop a stochastic model for the location of users. We show that the effect of interference on the performance degradation from the users with large propagation distance to their connecting satellites is a dominant factor; hence, to improve the performance of the system, we propose a method in which the transmissions of packets are controlled according to their distances to connecting satellites as well as traffic distribution.

From Section 7.2 to the end of the chapter, we discuss different issues in a LEO satellite system that have not been considered in detail in the preceding chapters. Those issues include the power control, its implementation, and the effects of its imperfections in a LEO satellite system; the design of LEO satellite systems with spot-beam antennas and the expression for antenna gain; and the new issue of adaptive array antennas in LEO satellite systems.

For the issue of power control, in Section 7.2, we explain that because of large propagation delay in a LEO satellite system, the closed-loop power control is much less effective than the one employed in digital terrestrial cellular systems. Moreover, the implementation of an accurate-enough power control in a LEO satellite system requires many more considerations than in a terrestrial system.

Section 7.3 explains the LEO satellite systems with multibeam antennas. The method is an important feature for a system with high capacity because it reuses the limited frequency spectrum and decreases multiple-access interference. In a LEO satellite system, however, design of effective spot beams requires many more considerations than terrestrial systems, especially the one that is the focus of this book, that is geographical nonuniformity in traffic distribution.

Section 7.5 introduces the issue of designing receivers with adaptive array antennas, which has been recently proposed for LEO satellite systems. This method can reduce the interference from undesired signals; however, in LEO satellite systems employed in the future PCNs with small hand-held terminals and wideband spread-spectrum signals, there would be some problems, which also are discussed in this section.

7.1 Packet admission control scheme

The preceding chapters introduced several methods for improving the performance of LEO satellite systems, both in uniform and nonuniform traffic situations. Chapter 6 proposed the family of TPC schemes, in which permissions for transmission of packets are controlled according to the propagation distance between users and their nearest satellites. This section proposes another method for improving the throughput performance of these systems, *packet admission control* (PAC), which has some similarities with the TPC scheme. In the PAC method, transmissions of packets are controlled according to the traffic distribution in addition to the distance of users to their connecting satellites. In this section, we modify our traffic model to a three-dimensional one and discuss the performance of the satellite system under heavy traffic situations and the effect of multiple-access interference on its throughput performance. Although there are some similarities between the system and the traffic models in this section and the ones used in previous chapters, we will explain the models briefly to highlight the differences.

7.1.1 System and traffic models

We are concerned with a multiorbit, multisatellite global communications network in which satellites are on LEOs of the altitude h. The number of orbits and the number of satellites on each orbit are designed so that any area on the globe is covered by at least one satellite at any given time. Users' terminals have the capability of direct access with satellites in both uplink and downlink directions. That assumption is realistic in the case of LEO satellite systems, because the low altitude of the satellites (i.e., the low propagation distances between users and satellites) allows low-power, hand-held personal terminals. We also make the preliminary assumption that any user communicates with the satellite that requires the lowest transmitting power, to minimize the total power of interference on the channel. Note that in a nonfading situation, that assumption means equal-size service areas for all satellites.

To establish a connection between a user and a satellite, it is necessary for the user to have an elevation angle larger than a minimum value θ_{min} to that satellite. The minimum elevation angle provides an upper bound on the service area of each satellite, which will be determined for a satellite system according to the constellation parameters, such as the number of orbits, the number of satellites, and their altitudes, as well as the access method utilized in the system.

We consider the performance of the LEO satellite system on its uplinks, that is, from users to satellites, in heavy traffic situations. In this direction, the management of multiple access of a large number of users transmitting packets to a single satellite directly affects the performance of the system. On the other hand, on the downlinks, any satellite can manage the flow of messages to users carefully in any traffic situations, by some conventional scheme such as TDMA or FDMA. If we consider the performance of uplinks in such a satellite system in which spread spectrum is employed for multiple accessing of users, the signal of any user located in the view of a satellite can be reached at that satellite and added to level of multiple-access interference at that satellite. Therefore, for any satellite, we should consider an interference area, the radius of which is defined by the elevation angle of zero.

We assume that our LEO satellite communications system operates under heavy traffic conditions and that a perfect power control mechanism is employed on uplinks. The system is slotted, and the packet lengths are assumed to be equal to the size of the slots. It also is assumed that the slot duration is sufficiently large to allow a preamble for spreading code and carrier synchronization. In each time slot, each user transmits a packet with a composite probability of q; that is, the probability of transmitting new packets and retransmitting failed packets from past slots. We are interested in calculating the performance of the system over many different changing topologies rather than for a specific terminal configuration. We especially are interested in examining the effect of interference from the terminals located inside the service area of a given satellite and the ones located outside that service area. As a result, we obtain statistical performance values over a set of topologies. To do that, we model the location of any user by a two-dimensional random variable that has a polar angle uniformly distributed between $[0, 2\pi)$ and the probability density of its distance to the origin has a normal-shape function of

$$f_R(r_u) = \frac{A}{\omega} \exp\left(-r_u^2/2\omega^2\right) \qquad (7.1)$$

where r_u is the distance of a user from the point that has the highest probability for the existence of users, which we refer to as the origin; ω is a parameter that shows how dense the terminals are distributed around the origin; and A is a parameter that makes the total probability of the existence of a user in the area under consideration equal to 1. The area under consideration here is a circle centered at the origin whose size is equal to the size of the interference area of a single satellite. With the assumption of equal-size service areas of satellites, the actual service area of each satellite has a near-hexagonal shape. Therefore, we define the equivalent service area as a hexagon with the center of the origin and the size of service area of a single satellite. We will evaluate the performance of the satellite system by counting the average number of packets successfully received at the satellite(s) covering the equivalent service area during a given period of time and refer to that measure as throughput.

Figure 7.1 shows the configuration of the service areas of the satellites at the instant when the service area of a satellite completely covers the equivalent service area. Note that since LEO satellites are in continuous motions, the situation shown in the figure is only an instantaneous configuration. Also note that because of the spherical shape of the globe,

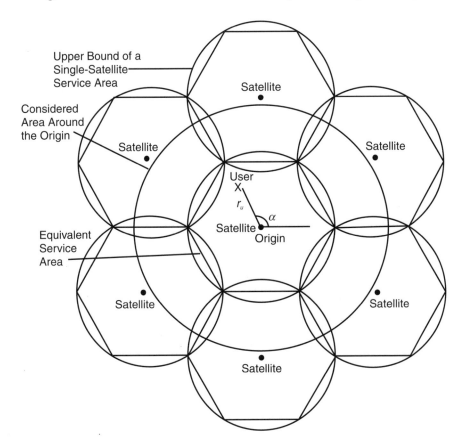

Figure 7.1   Configuration of service areas in the LEO satellite system.

to calculate the interference power from the terminals located outside the service area of a given satellite, it is enough to consider some portions of users located in the service areas of immediate neighbors. For example, at the instant shown in Figure 7.1, it is necessary to consider only the power of interference from users in the area indicated as "considered area around the origin." That area is equal in size to the interference of a single satellite. Then, for the satellite constellation parameters used in Subsection 7.1.2, for example, the size of this area will about 1.7 times the service area of a single satellite.

With this model, we can compare the effects of interference of the users when they are either inside or outside the equivalent service area on the throughput performance. After that comparison, we propose our method for improving the throughput performance.

7.1.2   Evaluation of heavy-traffic performance

This section shows the effect of multiple-access interference on the throughput performance of our satellite system. Throughout the following examples, a typical LEO satellite system with 6 orbits and 11 satellites on each orbit is considered. The orbit height, h, is 800 km, and the minimum elevation angle, θ_{min}, that users can connect to the satellites is 10. The total number of users, N_u, in the area under consideration is 100. Because we are interested in comparing the effect of interferences when their sources are inside or outside the equivalent service area, we use ω, defined in (7.1), as a parameter in the numerical examples. In most examples, we use typical values of 0.2, 0.5, and 0.75, which denote nonuniform high traffic situations; and ∞, which shows a uniform traffic situation.

Figures 7.2 and 7.3 show the normalized throughput, ξ_{norm}, which is the normalized value of the throughput by the expected number of users in the equivalent service area, of an unfaded channel as a function of composite transmission probability[1], q with ω as a parameter, for two values of the multiple-access capability $K(\mu_c) = 20$ and 30 (see Chapter 4 for more details on the throughput calculations). In the figures, the throughput characteristics are shown for two cases that the interferences

---

1. Note that the composite packet transmission probability q can be translated as the fraction of the total number of users transmitting simultaneously.

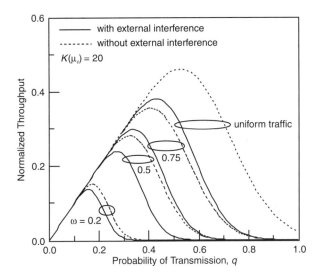

Figure 7.2 Comparison of throughput performances when the interferences from outside the equivalent service area are considered or ignored at different traffic situations of ω equal to 0.2, 0.5, 0.75, and ∞ (uniform traffic) with $K(\mu_c) = 20$.

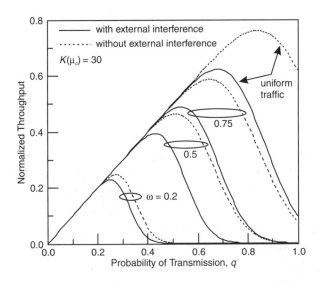

Figure 7.3 Comparison of throughput performances when the interferences from outside the equivalent service area are considered or ignored at different traffic situations of ω equal to 0.2, 0.5, 0.75, and ∞ (uniform traffic) with $K(\mu_c) = 30$.

from users outside the equivalent service area are either considered or ignored. Comparing the results of those two cases, we can see that the effect of interferences from surrounding service areas on the throughput performance degradation increases as the traffic distribution of users becomes more uniform. We can see that the largest difference in the performance of the system occurs in the uniform traffic distribution in which the number of users outside the equivalent service area reaches its maximum value.

In uniform traffic distribution, any user may be at any point inside the area under consideration with equal probability. However, in nonuniform traffic distributions, users mostly are located around the origin with high probabilities. Considering the evaluation period assumed in our analysis, that means that in nonuniform traffic distribution, we can expect a larger number of users with short propagation distances to their connecting satellites than in the uniform traffic situation. According to this discussion along with other numerical comparisons not shown here, we have confirmed that the interferences from users large distances from their connecting satellites have the dominant effect on the throughput performance in LEO satellite systems.

The other fact drawn from the two figures is the low throughput in the equivalent service area in the nonuniform traffic situation due to high multiple-access interferences. In that situation, most of the time, large numbers of packets are transmitted to a single satellite; hence, their probabilities of success decrease significantly.

For a LEO satellite system in which the effect of shadowing is also considered, different and interesting results can be derived. Shadowing of signals from users to satellites, caused by obstacles in paths between users and satellites, reduces the power of the received signal over the total signal bandwidth and its effects become larger as the elevation angle decreases. Because the reduction in power occurs not only for the tagged packet but also for interferers, in the case of a large number of interferers with low elevation angles, shadowing may decrease the power of interference and hence improve the throughput performance. To examine the effects of shadowing, we apply the fading model of [1], which was introduced in Section 6.2 with $c = 10$ as a typical value for LEO satellite channels and $B = 0.3$ as a compromised value between large values (for an urban area) and small values (for a suburban or rural area). Figure 7.4

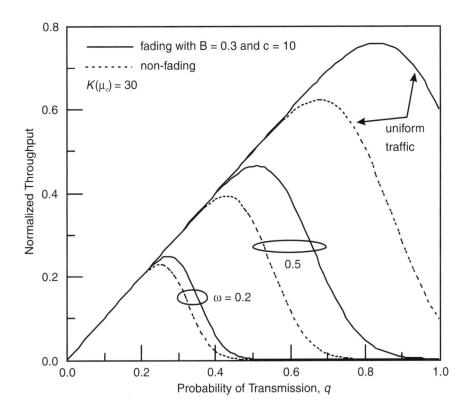

Figure 7.4 Comparison of throughput performances of fading and nonfading satellite channels at different traffic situations of ω equal to 0.2, 0.5, and ∞ (uniform traffic) with $K(\mu_c) = 30$.

compares the performance of a system in a nonfading channel to that in a faded channel for different traffic situations. As shown in the figure, the throughput performance in a fading channel is better than that of the nonfading one. The difference becomes larger when the traffic is more uniform. In a uniform traffic situation, the number of interferers with low elevation angles is larger than the number of those with nonuniform traffic distribution, which means lower power of interference and higher throughput. Because shadowing significantly reduces the power of interferences from users with low elevation angles, we can expect close throughput performances in the cases of shadowing assumption and ignore the external interference, as seen in Figures 7.3 and 7.4.

7.1.3   Concepts of the scheme

It has been shown that the power of the interference from users outside the equivalent service area has significant effect on degradation of the throughput performance of the system. Therefore, to improve the throughput performance of the system, we should reduce the interference effects. In Chapter 6, we proposed a TPC scheme in which the number of users with large propagation distances to their connecting satellites is decreased by prohibiting their transmissions until their propagation losses to connecting satellites fall under some threshold value. With the TPC, users of a satellite communication system at any instant are divided into two groups: permitted and unpermitted. With that method, we showed that the throughput performance of the system can be improved significantly. The TPC scheme makes use of the mobility nature of LEO satellites, which means the change of the permission condition for each user according to time.

This section proposes another scheme, PAC [2,3], in which, unlike the TPC, there is no prespecified unpermitted area, but the following two constraints are applied:

- Users with shorter propagation distances to their connecting satellites are honored with higher probability for transmission.

- The degree of priority for transmission of closer users over distant users is moreover controlled according to the traffic distribution of users in the area under consideration.

The first constraint aims to provide a higher probability of transmission to the users who require low-power transmitting signals, to decrease the total power of interference at each satellite. On the other hand, the second constraint is necessary to control the degree of priority given to the closer users by the first constraint. In the absence of the second constraint, even with a small number of users with short propagation distances, their priorities for transmissions are high. The second constraint biases those priorities to ensure more transmission from areas with more users in the case of nonuniform distribution of users.

There should be several approaches consistent with those two constraints. For an example of the PAC method that provides those two constraints, assume that in each slot each user may transmit a packet with probability $q\varepsilon$ instead of the simple probability of transmission, q, considered before. The parameter ε is a random variable between $[0, 1]$ with the probability density function $f_E(\varepsilon)$, which calculates by each user at the time of transmitting a packet according to traffic information provided by satellites on downlinks. A simple form of the random variable ε that supports the two constraints of the PAC is a linear function of the location of users, r_u, which is itself a random variable, such as

$$\varepsilon(r_u) = 1 - \frac{r_u}{r_0} \tag{7.2}$$

where r_0 is the radius of the equivalent service area and $r_u \leq r_0$. From (7.2), the probability distribution function of ε can be determined from that of the random variable r_u, in the form of

$$F_E(\varepsilon) = P[E \leq \varepsilon] = F_R[r_0(1 - \varepsilon)] \tag{7.3}$$

The linear relation of ε to r_u in (7.2) provides a higher probability of transmission for closer users in uniform traffic situations (i.e., the first constraint) because of capture of the traffic by the close satellite. Moreover, in nonuniform traffic distribution, larger values of ε are given to users located in the dense traffic areas.

7.1.4 Performance of the scheme

Let us now evaluate the throughput performance with the random packet transmission of the PAC method defined in Subsection 7.1.3 by numerical examples. In these examples, we assume the same parameters as those used in the examples in Subsection 7.1.2. Figures 7.5 and 7.6 show the normalized throughput as a function of the composite packet transmis-

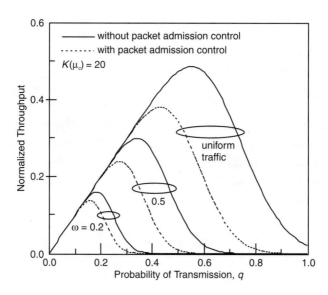

Figure 7.5 Throughput performance of PAC scheme at different traffic situations of ω equals to 0.2, 0.5, and ∞ (uniform traffic) with $K(\mu_c) = 20$.

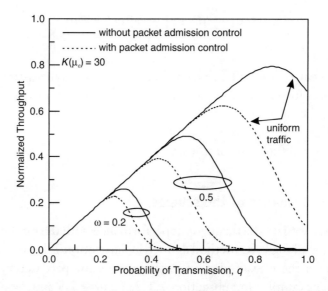

Figure 7.6 Throughput performance of PAC scheme at different traffic situations of ω equals to 0.2, 0.5, and ∞ (uniform traffic) with $K(\mu_c) = 30$.

sion, q, with ω as a parameter, for $K(\mu_c) = 20$ and 30, respectively, in an unfading channel assumption. As shown in the figures, assignment of higher packet admission probabilities to users with short propagation distances significantly improves the throughput characteristics in all traffic situations. In the case of $\omega = 0.2$, however, the performance improvement is rather small, since in that highly nonuniform traffic situation, the number of users with almost the same propagation distances to their connecting satellites at any time is large. It is interesting to note that by employing the PAC method, the throughput performances approach the ones found by the assumption in Section 7.1.2 of no interference from users outside the equivalent service area. Realization of such an assumption requires sharp antenna beams to reject interference from undesired areas, but with the PAC method, we can realize the same situation by soft rejection of interferences.

One may be interested in a comparison between the performance of the TPC and PAC schemes or in the performance of the PAC scheme under fading situation. For the former one, we should note that the TPC aims to improve the performance of the system, not of individual satellites, in nonuniform traffic situations. With the PAC, however, we can improve or maintain the performance of each satellite in different traffic distributions. Therefore, an exact comparison between the two schemes is a complicated task that requires further analyses. Regarding the performance of the PAC under fading conditions, for the LEO satellite channels it requires a more specific model than the one presented here. However, as the PAC decreases the number of users with low elevation angles, it will exhibit better performance under fading situation.

Considering the high ground speed of LEO satellites, the higher transmission priorities given to users with short propagation distances to their connecting satellites in the PAC method would not significantly affect the delay performance of the system. That is because of the higher probability of packet success realized with the PAC, which decreases the average number of retransmissions for each user. Moreover, because we are considered a slotted system, any increase in delay due to the PAC method may be included in the waiting time until the start point of the next slot, similar to the results shown in Chapter 6 for the case of TPC methods.

7.2　Power control

7.2.1　The near-far problem

In digital terrestrial cellular systems employing CDMA, the most serious difficulty that directly affects the total capacity of the system, is the near-far problem. In such a multiuser CDMA system, all users are contending for the same bandwidth at the same time, and then users interfere with each other, which is referred to as the multiple-access interference effect. In a wireless communication system, the propagation loss law implies that the received power decreases as the distance between transmitter and receiver increases, which can be described by the following relation:

$$P_r \approx d^{-\beta} P_t \tag{7.4}$$

where P_r and P_t are the received and the transmitted powers, respectively; d is the distance between receiver and transmitter; and β is the propagation loss factor. In a system with almost direct line of sight, such as a satellite system, the value of β is about 2; however, in terrestrial cellular systems, the value of β is in the range of 2 to 5. In the latter case, the radio waves are reflected or partially absorbed by obstacles between receiver and transmitter as well as the surface of the Earth; hence, larger values for β result.

Due to the propagation loss law, if two users, one close to the base station and the other far from the base station, transmit with equal power levels, the received power at the base station from the closer user will always be much stronger than that of the more distant user. Therefore, signals from the more distant user are masked by interference from the closer user. This problem is referred to as the near-far problem and is the most important problem facing CDMA systems. To maximize the capacity of such systems, system resources should be shared equitably among users, which means we should provide an appropriate solution for the near-far problem.

There are a number of methods to solve the near-far problem in CDMA systems, perhaps chief among them the well-known power control method [4–15]. The power control on reverse links, which we used as a basic assumption in preceding chapters, is a method with which we can equalize the received power from all transmitters at the base station. The most serious problem facing power control is the realization of a perfect power control that can result in equal-level power from all transmitters at the receiver in all situations (e.g., fading channels) is difficult, even impossible. As we show soon, imperfections in power control in a CDMA system significantly decrease the capacity of the system.

Because the near-far problem occurs mostly in CDMA systems, which employ correlation receivers to solve the problem, designs of receivers other than the correlation receiver have been proposed [16–22]. Those methods consider different interference cancellation algorithms, such as parallel or successive, and receivers with decorrelators. We do not discuss these methods in this book and suggest the reader see the references for more details. The following subsections focus the discussion on the power control method and the effects of imperfections in that method.

7.2.2 Implementation of power control

Power control is the most well-known method for alleviating the near-far problem in CDMA systems with correlation receivers. Generally, power control should be designed in both uplinks and downlinks in such systems to avoid the near-far problem. In general, there are two methods of power control: the open loop and the closed loop. Considering the power control on uplinks from users to base stations, in the open loop each user adjusts the transmit power based on the level of received power from the base station. In the closed loop, the base station commands users to either increase or decrease their power, which is based on the level of the signal received from each user.

One method of implementation of power control that has been reported in terrestrial cellular systems is the use of pilot signals transmitted by base stations [8]. In those systems, the base station sends a pilot signal with a nominal level, and each user monitors the power of that signal. According to that power level, the user transmits a signal at an initial power level corresponding to the nominal level of the pilot signal and the received level. In a cellular system with relatively large cells, it is reported that a user should be capable of controlling transmitting power as much as 80 dB in some situations [8].

There are two significant difficulties in the implementation of that simple method. The first one, which is common in both terrestrial and satellite systems, is that the condition of channels in downlinks and uplinks may not be the same. That is, the pilot signal on the downlink experiences different fading variations than that of the transmitting signal from the user on uplinks. Moreover, the fading on the uplink usually is uncorrelated with that on the downlink, since the uplink and downlink frequency bands usually are separated by more than the coherence bandwidth of the channel. The second problem, which is more critical in the case of satellite systems, is the duration of the power control process. The variations due to Rayleigh fading usually are too rapid to be tracked by the closed-loop power control. That is particularly almost impossible when we consider LEO satellite systems with a round trip much larger than that in a terrestrial system, on the order of 10 ms for a 780-km altitude to 60 ms for a 1,500-km altitude at lower elevation angles. The variations in relative path losses and shadowing effects, however, generally are slow enough to be controlled when they are modeled as an attenuation with log-normal distribution. Therefore, for LEO satellite systems, closed-loop power control is much less effective than it is on a terrestrial channel. When closed-loop power control is not possible for a LEO satellite system, achieving an acceptable accuracy level of power control based on an open-loop approach requires much more work.

Another significant difference between a terrestrial channel and a LEO satellite is that the multipath delay spread in a LEO satellite channel is more than that of a terrestrial channel, on the order of 100 ns. Therefore, the coherence bandwidth of the satellite multipath channel is at least 10 MHz, which means that, in a CDMA design, mitigating the multipath spreading by an amount greater than 10 MHz is required in the case of LEO satellite systems [30].

7.2.3   Effects of imperfections in power control

Imperfection in power control and its effect on the capacity of a system have been widely studied in CDMA-based terrestrial systems [5,23–29].

In the case of LEO satellite systems, there are several reports in the literature [30,31]. In practice, the power control exhibits some imperfections that result in different received powers from users at the base station. The imperfections in power control can be measured by the logarithmic standard deviation of the received power, P_r, σ, which in the case of perfect power control is equal to zero. Usually the received power is described by a log-normal distributed random variable in the form of [25,26]

$$f(P_r) = \frac{1}{\sqrt{2\pi}\ \sigma P_r} \exp\left[-\frac{\ln(P_r)}{2\sigma^2}\right] \qquad (7.5)$$

In [26], it is reported that for the reverse link of a slotted direct sequence CDMA cellular system with a processing gain of 255 and 200 users per cell, the maximum value of throughput[2] of each cell, defined as the average number of successfully received packets per time slot, is about 28 for a perfect power control system but decreases to 18 and 13 due to imperfection in power control of $\sigma = 2$ dB and $\sigma = 4$ dB, respectively. Moreover, it is shown that the delay performance of a cellular system degrades significantly as the power control error increases.

In the case of LEO satellite systems, usually a time-shared mixture density function is used to model the multipath fading and shadowing channel. That model has been proposed in [1] and was explained in Chapter 6. In the model, it is assumed that in the shadowed fraction of time, B, the channel is log-normal Rayleigh fading, whereas in the unshadowed period, $1-B$, it is Rician. Then, for a land-mobile satellite channel, the probability density function of the received amplitude can be described by

$$f_R(R) = 2BRce^{-cR^2} + 2(1-B)Rce^{-c(R^2+1)} I_0(2Rc) \qquad (7.6)$$

where $c = A_s/2\sigma_s^2$, A_s is the amplitude of the specular component of the Rician part of the density, $2\sigma_s^2$ is the average power in the scatter

---

2. Maximum value of throughput in a packet communication system is usually called the capacity of the system.

component of the fade, and $I_0(\cdot)$ is the modified Bessel function of the first kind and zeroth order. Borrowing the notations used in [30], if we denote the nominal received power in the absence of either fading or shadowing by S_{nom}, then in the absence of shadowing but in the presence of Rician fade, the average received power is

$$S_{av_{ns}} = S_{norm}\left(1 + \frac{1}{c}\right) \qquad (7.7)$$

If the user is shadowed, the received power is

$$S_{av_s} = S_{norm}\left(\frac{1}{c}\right) \qquad (7.8)$$

Usually the power control cannot track the power variations due to multipath fading, but it can track the variations due to the shadowing. Then, during the shadowing period, each user multiplies transmitted power by

$$p = \frac{S_{av_{ns}}}{S_{av_s}} = 1 + c \qquad (7.9)$$

Vojcic et al. have given a complete analysis for the bit error rate for a direct sequence CDMA in a LEO satellite system in which the footprint of each satellite is partitioned by spot beams considering both uncoded and convolutional coded CDMA system [30]. In their paper, they reported that with a processing gain of 150 of an uncoded CDMA and typical values of $c = 10$ and $B = 0.3$, even with relatively small power control error on the order of 0.5-dB standard deviation, there is a large difference between the number of users that can simultaneously be active in each spot beam in shadowed and unshadowed conditions. Even with that small power control error, users of the satellite system cannot achieve the bit error rate of 10^{-3} in $E_b/N_0 = 30$ dB. It is also shown that the performance of unshadowed users is more sensitive to variations of power control error, because the performance of shadowed users is dominated by the effect of fading.

In the case of a CDMA system with a convolutional code of the rate $1/3$ and constraint length 8 with maximum likelihood decoding in the receiver, it is shown in [30] that with a single satellite visible to users at any time, only 10 users can be supported with a bit error rate of 10^{-3} in 2-dB standard deviation of power control error. If under the same conditions more accurate power control is employed so that a standard deviation of 0.5 is available, then about 35 users can be supported in each spot beam. This shows that the capacity of the LEO satellite systems, that is, the number of simultaneous users, is very much affected by the degree of imperfection in power control. Vojcic et al. concluded that the direct-sequence CDMA will be a viable multiple-access technique in LEO satellite systems if sufficient interleaving, relatively good power control methods that provide a standard deviation in power control error less than 2 dB, and dual diversity that is visible to more than one satellite per user at all times can be employed.

We should mention that in the case of LEO satellite systems much more care should be taken during the implementation of power control than is required in terrestrial cellular systems. The satellite channel has more fading variations than that of a terrestrial system, and the accuracy in power control significantly affects the capacity of the system. If a perfect power control is too expensive or too difficult to implement in CDMA-based LEO satellite systems, then the design of receivers based on other methods than the correlation receiver would be a better choice.

7.3 Multibeam LEO satellites

As introduced in Chapter 2, many proposals for the LEO satellite system consider multiple spot beams, which partition the footprints of satellites into smaller cells. On the other hand, all the analyses presented in this book consider single-beam antennas. The question is how the partitioning of the footprints into smaller cells affects the results. This section presents a brief discussion of the issue of antenna gain; after that, we examine different aspects of multibeam satellite systems.

7.3.1   General expression for antenna gain

Evaluating the change in the performance of a satellite system due to partitioning the footprints of satellites by spot beams directly requires knowledge of antenna theory. Although the aim of this book is not to discuss the issue of antenna design, we present it here briefly.

The *gain* of an antenna is defined as the ratio of the power per unit solid angle radiated in a given direction from the antenna to the power per unit solid angle radiated from an isotropic antenna supplied with the same power [32]. The isotropic antenna is a hypothetical antenna that radiates radio waves of a constant strength in every direction, or in 4π steradians. Such an antenna is used as the reference for microwave antenna gain; it cannot, however, be realized physically because it could not create transverse polarized electromagnetic waves. Assume the radiated field in the direction (α, ϕ) (α is the angle between the considered direction and the one in which maximum power is radiated, often called the *boresight;* ϕ is the phase) as $F(\alpha, \phi)$, then the gain of the antenna at the angle α, $G(\alpha, \phi)$, is given by

$$G(\alpha, \phi) = \frac{F(\alpha, \phi)}{P_0/4\pi} \tag{7.10}$$

where P_0 is the total power radiated by the test antenna. If the total radiated power of a transmitting source is P_t, then

$$D(\alpha, \phi) = \frac{F(\alpha, \phi)}{P_t/4\pi} \tag{7.11}$$

expresses the *directivity*, and

$$\frac{G(\alpha, \phi)}{D(\alpha, \phi)} = \frac{P_t}{P_0} = \eta \tag{7.12}$$

expresses the *aperture efficiency*.

In general, the value $G(\alpha_0, \phi_0)$ in (7.10) in the direction (α_0, ϕ_0) where a maximum radiation occurs is simply called the gain. Although there may be more than one direction in which the maximum radiation occurs, usually the gain of an antenna is defined in the direction of the *boresight* of the antenna, that is, the value of G at angle $\alpha = 0$ [33]. The gain of an antenna having the physical aperture area of A is expressed by

$$G = \frac{4\pi}{\lambda^2} A\eta \tag{7.13}$$

where λ is the wavelength. The coefficient η is the aperture efficiency; therefore, $A\eta$ represents the effective aperture area. If the aperture is a disk of diameter D, then (7.13) can be written as

$$G = \left(\frac{\pi D}{\lambda}\right)^2 \eta \tag{7.14}$$

The gain is normally expressed by a value in decibels in reference to the isotropic antenna, that is,

$$G = 10\log_{10}\eta\left(\frac{\pi D}{\lambda}\right)^2 \tag{7.15}$$

The coefficient $4\pi/\lambda^2$ in (7.13) is called the *universal constant* [34], which universally lies between the gain and the effective area of all kinds of antennas. If $G = 1$ and $\eta = 1$, (7.13) becomes $A_0 = \lambda^2/4\pi$. Therefore, A_0 can be interpreted as the effective area of the isotropic antenna; then it is called the *isotropic area*. The product of $P_t G_t$, where G_t is the gain of a lossless antenna, often is called the *effective isotropically radiated power* (EIRP) and describes the combination of the transmitter and the antenna in terms of an equivalent isotropic source with power $P_t G_t$, in watts, radiating uniformly in all directions.

7.3.2 Spot-beam antenna gain

This subsection discusses the antenna gain in a LEO satellite system with spot-beam antennas. Assume that the total number of satellites in a LEO constellation, that is, the multiplication of the total number of satellites in each orbit and the number of orbits, is N_{ts} and the footprint of each satellite is partitioned into N_c cells. Then, any specified cell on the Earth can be identified by the couple (j, k), where $j = 1, \ldots, N_c$ and $k = 1, \ldots, N_{ts}$.

Each cell is covered by the main lobe of a spot beam whose gain, $G_j(\alpha)$, is related to the normalized far-field radiation pattern, $F_j(\alpha)$, by [35]

$$G_j(\alpha) = G_{Mj} \, F_j^2(\alpha) \qquad (7.16)$$

where α is the angle between the main (central) beam direction or boresight (see Figure 7.7) and G_{Mj} is the maximum gain of the jth spot. Each spot beam may have different patterns to compensate for the different angles of incidence and free-space losses. By modeling the radiation patterns through suitable masks enveloping the maxima of the generic tapered-aperture antenna radiation pattern, $F_j(\alpha)$ would be [35]

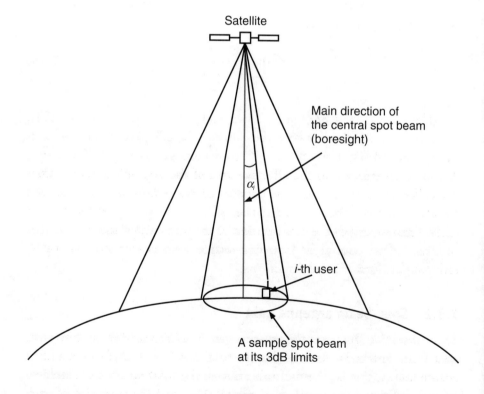

Figure 7.7 Illustration of different definitions in spot beam configuration.

$$F_j(\alpha) = \hat{F}(u_j) = \frac{(p+1)(1-T)}{(p+1)(1-T)+T} \cdot \left(\frac{2J_1(u_j)}{u_j} + 2^{p+1} p! \frac{T}{1-T} \frac{J_{p+1}(u_j)}{u_j^{p+1}} \right) \quad (7.17)$$

where $u_j = \pi d_{aj} \sin \theta / \lambda$, $J_p(u)$ is the Bessel function of the first kind and order p, T is the aperture edge taper, d_{aj} is the effective aperture diameter of the jth spot, and λ is the wavelength.

If d_{a1} denotes the effective aperture diameter for the spot covering the subsatellite point, d_{aj} can be evaluated as [35]

$$\frac{d_{aj}}{d_{a1}} = 1 + c \left[1 + \left(\frac{r_j}{r_{min}} \right)^{\varepsilon} \right] \quad (7.18)$$

where r_j is the distance between the jth spot boresight intersection on Earth and the subsatellite point; r_{min} is the minimum value for r_j; c is a constant equal to 0.05; and ε is an empirical function of the satellite altitude h; and the elevation angle, θ, in degrees defined as

$$\varepsilon = \varepsilon(h, \theta) = 3.39 - 0.0385\theta - 0.748\log_{10}(h/h_0), \quad \varepsilon \geq 1 \quad (7.19)$$

with $h_0 = 1,000$ km. Figure 7.8 is an example of the spot-beam radiation diagram, $\hat{F}(u)$, for $T = 20$ and $p = 2$ [35]. The presence of sidelobes seen in the radiation diagram results in cochannel interference between adjacent cells and implies nonideal angular selectivity of the spot beams. In [35], there are also some mathematical analyses on the level of carrier-to-interference ratio in the case of spot-beam antennas.

7.3.3 Performance of spot-beam antennas

Let us now discuss the effect of spot-beam antennas on the performance of the system. The idea of having multiple cells within the footprint of a satellite in a CDMA-based system is to increase the total capacity of the system by limiting the number of interfering users to those inside each cell. In that manner, the approach of spot-beam antennas can be assumed as an interference cancellation method, which is studied widely in the literature [35–46]. Generally, we can assume two kinds of frequency allocation in a multicell CDMA system. The first one is to assign the same

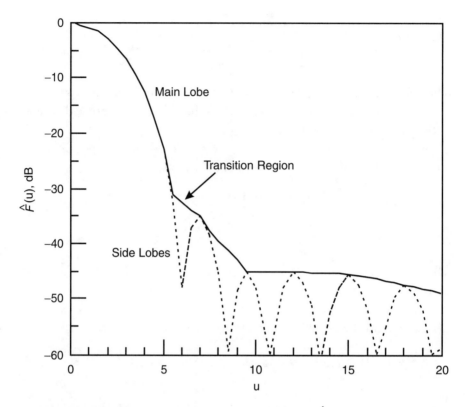

Figure 7.8 Normalized far-field radiation pattern $\hat{F}(u)$ of a typical spot beam for $T = 20$ dB and $p = 2$.

frequency band in all cells and different sets of orthogonal codes in different cells. In that method, the available number of orthogonal codes will be divided by the number of cells; hence, we have to maintain the number of simultaneous users in each cell within the number of codes. Obviously with this method, the hand-off process would be simpler, because the users do not have to change their frequency when leaving a cell and entering an adjacent cell, and soft hand-off becomes possible.

The second possibility is to divide the available frequency band into a number of subbands with some guard bands between each pair of adjacent subbands. The number is determined according to how much the cells with the same frequency bands should be separated to reuse the frequency band. For example, if a seven-frequency reuse pattern is used whereby the central cell is surrounded by six cells, we should divide the total frequency band by seven. With this method, it is possible to reuse the same code in adjacent cells, and then the total number of available codes can be used in each cell. Thus, larger number of users can access the CDMA channel simultaneously. Here, users have to change their frequency when changing their host cell.

Because the first method has a wider bandwidth and a smaller number of codes, and the second method has a narrower bandwidth and a larger number of codes, it should be possible to make a trade-off between the two methods. Usually a seven-frequency reuse pattern is used, although sometimes a three-frequency pattern is employed. In practical systems, it is assumed that signals of users are perfectly orthogonal within a beam, but signals of different beams are not orthogonal. Therefore, signals of other beams appear as pseudo-noise in the frequency reused beams, which their summation can be assumed as a Guassian noise.

Frequency reuse in a satellite channel depends critically on the spacecraft antenna, which is the primary source of isolation between users with the same frequency band. In particular, it depends on the number of spot beams and their size. In that manner, perhaps one spot beam for each user would be the ideal case, although it is impractical. Actually, a study presented in [41] has shown that in either a narrowband or a wideband satellite system increasing the number of spot beams to more than 37 in a three-frequency reuse pattern only slightly changes the interference distribution. Therefore, in a real situation, there is an asymptotic gain due to having spot-beam antennas.

In the case of GEO satellite systems that employ spot-beam antennas, perhaps the simplest configuration is the one in which spot beams are pointed permanently in a given direction. However, Reudink et al. have shown that, in certain situations, scanning spot beams can make optimal use of satellite power [39]. In particular, in areas where there is low traffic density, dedicated spot beams would be underutilized and thus represent uneconomical satellite design. In such cases, scanning spot beams could be used to advantage. A side benefit would be the reduction of interference because at any time there would be fewer beams than service areas. A similar conclusion would be true for satellite systems in low Earth orbits.

From the preceding discussions along with the explanations given in Chapter 2, we conclude that the issue of multibeam antennas for the LEO satellite system is a serious requirement for future PCNs, to increase their total capacity. However, their implementation in LEO satellite systems requires many considerations, more than the ones usually considered in GEO satellite systems. Perhaps the most important consideration is the traffic distribution of users in the system. An effective LEO satellite system may be the one that has intelligent-enough on-board satellite processors that can manage the system resources, that is, the available bandwidth and codes, in such a manner that more resources are given to the areas with higher traffic loads without wasting them in areas with low traffic. The bounds of the intelligence of the system would be wide. Examples of an effective LEO satellite system would be a system with the ability to change the number of beams per satellite; a system in which the size of every spot beam can be changed according to the traffic load of the area in which it is projected; a system with the capability of assigning a different number of codes in different spot beams; a system that can change the number of frequency reuse patterns between spot beams; a system that can assign different frequency bandwidths within the available spectrum to different cells according to the requirements of the quality of service in those cells.

7.4 Concept of adaptive array antennas

It should be clear by now that in a LEO satellite system, the satellites have a high relative ground speed, and the position of any satellite relative to a user on the ground changes rapidly even during a short transmission. A similar situation can be seen in terrestrial mobile communication system, but with much lower speed. Moreover, there is a necessity to control the level of interference from users in the service area of other satellites or in a spot-beam constellation from users in other cells to achieve an acceptable level of signal-to-noise ratio.

One solution to those issues is the employment of adaptive array antennas at the users' terminals[3], which has been proposed for terrestrial cellular systems [47–54] and for LEO satellite systems [55]. An adaptive array antenna is an array of antenna elements whose overall directivity pattern is controlled automatically so the reception is made under the

---

3. Having adaptive array antennas at the base stations also has been proposed.

best signal-to-noise plus interference ratio. Because an adaptive array antenna adaptively controls the directivity of the antenna, it can be useful in suppressing interfering signals in a spread-spectrum system. Note, however, that an array antenna cannot suppress a high-power interfering signal from an undesired user. In that situation, perhaps a combination of adaptive array antennas with other methods of interference cancellation would be a better choice [51].

The typical structure of a conventional adaptive array antenna with N elements physically separated by the spaces L_i, $i = 2, \ldots, N$, is shown in Figure 7.9 [51]. In that structure, each complex-valued received signal X_k, $k = 1, 2, \ldots, N$, is composed of multipath signals and thermal noise and is a function of time. The received signals, after amplification, for example, are multiplied by appropriate weights. The reference signal is necessary to adjust the weights and ideally is a desired signal. In a direct-sequence CDMA system, for example, one can assume the transmission of a training signal whose code pattern is known at the receiver for a short period at the beginning of the communication. Then, for the training period, a carrier at the receiver is modulated by the training code and the reference signal is generated.

With a good decision on the complex values of the weights, W_k, the adaptive array yields automatic beam tracking of the desired signal and adequate suppression of interferer signals. Therefore, the desired signal can be amplified, while other undesired signals are attenuated. One problem with this receiver is that if the arrival angle of an interfering signal in a wideband system is the same as that of the desired signal, it cannot suppress the interfering signal. A complete rejection of interfering signals with this method would be available only for narrowband systems. Therefore, other interference-cancellation methods should be used in cooperation with adaptive array antennas in wideband spread-spectrum systems. Another issue related to receivers based on array antennas is the physical spacing between the elements of the antenna. Although we have almost no problem with spacing the elements wide enough in a large ground station, for a hand-held receiver of a mobile user spacing would be limited.

As a closing comment on the issue of employing adaptive array antennas in LEO satellite systems, we can say that the design of a LEO satellite system with intelligent cell configuration, adaptive array an-

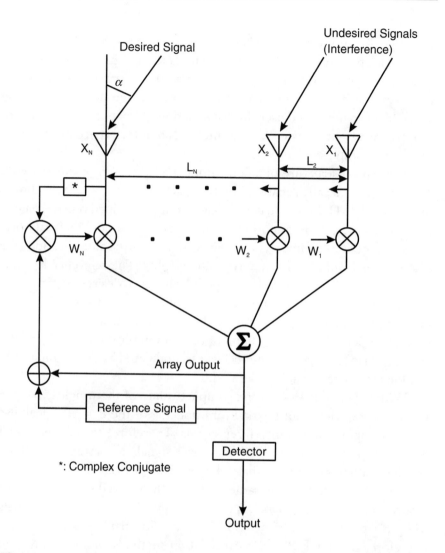

Figure 7.9   A conventional adaptive array antenna with N elements.

tenna-based receivers, and multiple diversity, that is, the visibility of more than one satellite (or being in more than one spot beam) for a given user at any time, can reduce the level of multiple-access interference and hence improve the capacity of the system significantly. Of course, implementation of all these items requires higher total cost of the system, which needs an appropriate trade-off in the system design.

7.5 Summary

This chapter introduced a method for improving the performance of LEO satellite systems in the case of spread-spectrum utilization, namely, the PAC scheme. Although this method has some similarities to the TPC method introduced in Chapter 6, the relative control of transmissions according to the traffic load in each area in the PAC method achieves much more interesting performance than the TPC method. That is because in the PAC method the probabilities of packet transmission are managed according to propagation distances of the originator of each packet as well as the traffic distribution. We have shown that such control improves the throughput characteristics of the satellite system in different traffic situations significantly.

We also discussed issues concerning a LEO satellite system that were not discussed in detail in previous chapters, such as power control, its implementation methods, and the effect of imperfections in power control; antenna gain; and multiple-spot-beam satellite systems. In addition, we introduced the issue of design of receivers with adaptive array antennas as a method for reducing the effect of interference, which has been recently considered in LEO satellite systems. We concluded that the accuracy in power control is a requirement for LEO satellite systems employing CDMA techniques and that it is difficult to design a perfect power control because of the relatively large propagation delay in a satellite system. Large delay makes it impossible to employ closed-loop power control in a LEO satellite system.

The discussion of LEO satellite systems using spot-beam antennas concluded that the issue of multibeam antennas for LEO satellite systems is a serious requirement in future PCNs to reuse the limited frequency spectrum and hence increase their total capacity. However, implementation of spot beams in LEO satellite systems requires considerations other than those usually considered in GEO satellite systems because of traffic distribution of users in the system. Perhaps a LEO satellite system with intelligent beam configuration, so the interference from other cells could be reduced to its minimum value, would be a final goal in design of LEO satellite systems.

For the issue of adaptive array antenna in LEO satellite systems, we concluded that although the design of receivers with array antenna is a good method for reducing interference and improving capacity, in a wideband system other interference-cancellation methods may be required as well. Moreover, physical spacing between the elements of the antenna in such a configuration might be a problem for the systems with small hand-held user terminals to be employed in the global PCNs of the next century.

References

[1] Lutz, E., et al., "The Land Mobile Satellite Communication Channel—Recording, Statistics and Channel Model," *IEEE Trans. Vehic. Technol.*, Vol. 40, No. 2, 1991, pp. 375–386.

[2] Jamalipour, A., and A. Ogawa, "Packet Admission Control in a Direct-Sequence Spread Spectrum LEO Satellite Communication Network," *IEEE J. Select. Areas Commun.*, Vol. 15, No. 8, 1997.

[3] Jamalipour, A., and A. Ogawa, "Packet Admission Control for Nonuniform Traffic Direct-Sequence Spread-Spectrum LEO Satellite Communications Networks," *Proc. Int. Conf. Telecommunications (ICT '97)*, Melbourne, Australia, 1997, pp. 1243–1248.

[4] Viterbi, A. J., *CDMA Principles of Spread Spectrum Communication*, Reading, MA: Addison-Wesley, 1995.

[5] Prasad, R., *CDMA for Wireless Personal Communications*, Norwood, MA: Artech House, 1996.

[6] Gilhousen, K. S., et al., "Increased Capacity Using CDMA for Mobile Satellite Communication," *IEEE J. Select. Areas Commun.*, Vol. 8, No. 4, 1990, pp. 503–514.

[7] Lee, W. C. Y., "Overview of Cellular CDMA," *IEEE Trans. Vehic. Technol.*, Vol. 40, No. 2, 1991, pp. 291–302.

[8] Gilhousen, K. S., et al., "On the Capacity of a Cellular CDMA System," *IEEE Trans. Vehic. Technol.*, Vol. 40, No. 2, 1991, pp. 303–312.

[9] Pickholtz, R., L. B. Milstein, and D. L. Schilling, "Spread Spectrum for Mobile Communications," *IEEE Trans. Vehic. Technol.*, Vol. 40, No. 2, 1991, pp. 313–322.

[10] Simpson, F., and J. Holtzman, "CDMA Power Control, Interleaving, and Coding," *Proc. 41th IEEE Vehic. Technol. Conf.*, St. Louis, 1991, pp. 362–367.

[11] Viterbi, A. J., A. M. Viterbi, and E. Zehavi, "Performance of Power-Controlled Wideband Terrestrial Digital Communication," *IEEE Trans. Commun.*, Vol. 41, No. 4, 1993, pp. 559–569.

[12] Viterbi, A. M., and A. J. Viterbi, "Erlang Capacity of a Power Controlled CDMA System," *IEEE J. Select. Areas Commun.*, Vol. 11, No. 6, 1993, pp. 892–900.

[13] Kchao, C., and G. L. Stuber, "Analysis of a Direct-Sequence Spread-Spectrum Cellular Radio System," *IEEE Trans. Commun.*, Vol. 41, No. 10, 1993, pp. 1507–1516.

[14] Viterbi, A. J., et al., "Soft Handoff Extends CDMA Cell Coverage and Increases Reverse Link Capacity," *IEEE J. Select. Areas Commun.*, Vol. 12, No. 8, 1994, pp. 1281–1288.

[15] Sheikh, A., Y. Yao, and S. Cheng, "Throughput Enhancement of Direct-Sequence Spread-Spectrum Packet Radio Networks by Adaptive Power Control," *IEEE Trans. Commun.*, Vol. 42, No. 2/3/4, 1994, pp. 884–890.

[16] Lupas, R., and S. Verdu, "Linear Multiuser Detectors for Synchronous Code-Division Multiple Access Channels," *IEEE Trans. Inform. Theory*, Vol. 35, No. 1, 1989, pp. 123–136.

[17] Lupas, R., and S. Verdu, "Near-Far Resistance of Multiuser Detectors in Asynchronous Channels," *IEEE Trans. Commun.*, Vol. 38, No. 4, 1990, pp. 496–508.

[18] Varanasi, M. K., and B. Aazhang, "Multistage Detection in Asynchronous Code-Division Multiple Access Communications," *IEEE Trans. Commun.*, Vol. 38, No. 4, 1990, pp. 509–519.

[19] Patel, P., and J. Holtzman, "Analysis of a Simple Successive Interference Cancellation Scheme in a DS/CDMA System," *IEEE J. Select. Areas Commun.*, Vol. 12, No. 5, 1994, pp. 796–807.

[20] Kohno, R., et al., "An Adaptive Canceller of Co-Channel Interference for Spread Spectrum Multiple Access Communication Networks in a Power Line," *IEEE J. Select. Areas Commun.*, Vol. 8, No. 4, 1990, pp. 691–699.

[21] Rush, L. A., and H. V. Poor, "Narrowband Interference Suppression in CDMA Spread Spectrum Communications," *IEEE Trans. Commun.*, Vol. 42, No. 4, 1994, pp. 1969–1979.

[22] Yoon, Y. C., R. Kohno, and H. Imai "A Spread-Spectrum Multi-Access System With Co-Channel Interference Cancellation Over Multipath Fading Channels," *IEEE J. Select. Areas Commun.*, Vol. 11, No. 7, 1993, pp. 1067–1075.

[23] Kudoh, E., and T. Matsumoto, "Effects of Power Control Error on the System User Capacity of DS/CDMA Cellular Mobile Radios," *IEICE Trans. Commun.*, Vol. E75-B, No. 6, 1992, pp. 524–529.

[24] Kudoh, E., "On the Capacity of DS/CDMA Cellular Mobile Radios Under Imperfect Transmitter Power Control," *IEICE Trans. Commun.*, Vol. E76-B, No. 8, 1993, pp. 886–893.

[25] Prasad, R., M. G. Jansen, and A. Kegel, "Capacity Analysis of a Cellular Direct Sequence Code Division Multiple Access System With Imperfect Power Control," *IEICE Trans. Commun.,* Vol. E76-B, No. 8, 1993, pp. 894–905.

[26] Jansen, M. G., and R. Prasad, "Throughput and Delay Analysis of a Cellular Slotted DS CDMA System With Imperfect Power Control and Sectorization," *Proc. IEEE International Symposium on Spread Spectrum and Its Applications (ISSSTA '94),* Oulu, Finland, 1994, pp. 420–425.

[27] Newson, P., and M. R. Heath, "The Capacity of a Spread Spectrum CDMA System for Cellular Mobile Radio With Consideration of System Imperfections," *IEEE J. Select. Areas Commun.,* Vol. 12, No. 4, 1994, pp. 673–684.

[28] Cameron, R., and B. Woerner, "Performance Analysis of CDMA With Imperfect Power Control," *IEEE Trans. Commun.,* Vol. 44, No. 7, 1996, pp. 777–781.

[29] Jacobsmeyer, J. M., "Congestion Relief on Power-Controlled CDMA Networks," *IEEE J. Select. Areas Commun.,* Vol. 14, No. 9, 1996, pp. 1758–1761.

[30] Vojcic, B. R., R. L. Pickholtz, and L. B. Milstein, "Performance of DS-CDMA With Imperfect Power Control Operating Over a Low Earth Orbiting Satellite Link," *IEEE J. Select. Areas Commun.,* Vol. 12, No. 4, 1994, pp. 560–567.

[31] Monk, A. M., and L. B. Milstein, "Open-Loop Power Control Error in a Land Mobile Satellite System," *IEEE J. Select. Areas Commun.,* Vol. 13, No. 2, 1995, pp. 205–212.

[32] Silver, S., *Antenna Theory and Design,* New York: McGraw-Hill, 1949.

[33] Pratt, T., and C. W. Bostian, *Satellite Communications,* New York: Wiley & Sons, 1986.

[34] Slater, J. C., *Microwave Transmission,* New York: McGraw-Hill, 1942.

[35] Vatalaro, F., et al., "Analysis of LEO, MEO, and GEO Global Mobile Satellite Systems in the Presence of Interference and Fading," *IEEE J. Select. Areas Commun.,* Vol. 13, No. 3, 1995, pp. 291–300.

[36] Roddy, D., *Satellite Communications,* Englewood Cliffs, NJ: Prentice-Hall, 1989.

[37] Williamson, M., *The Communications Satellite,* Bristol and New York: Adam Hilger, 1990.

[38] Jansky, D. M., and M. C. Jeruchim, *Communication Satellites in the Geostationary Orbit,* Norwood, MA: Artech House, 1987.

[39] Reudink, D. O., A. S. Acampora, and Y. S. Yeh, "The Transmission Capacity of Multibeam Communication Satellites," *Proc. IEEE,* Vol. 69, No. 2, 1981, pp. 209–225.

[40] Gaaudenzi, R., et al., "How Can Interference-Rejection Receivers Increase the Capacity of CDMA Multi-Beam Satellite Communication Systems?," in *Mobile and Personal Satellite Communications 2,* F. Vatalaro and F. Ananasso, eds, London: Springer, 1996, pp. 349–365.

[41] Moher, M., et al., "Interference Statistics for Multibeam Satellites," in *Mobile and Personal Satellite Communications 2,* F. Vatalaro and F. Ananasso, eds, London: Springer, 1996, pp. 366–384.

[42] Ananasso, F., and F. D. Priscoli, "The Role of Satellites in Personal Communication Services," *IEEE J. Select. Areas Commun.,* Vol. 13, No. 2, 1995, pp. 180–195.

[43] Vojcic, B. R., L. B. Milstein, and R. L. Pickholtz, "Total Capacity in a Shared CDMA LEOS Environment," *IEEE J. Select. Areas Commun.,* Vol. 13, No. 2, 1995, pp. 232–244.

[44] Kim, J. Y., and J. H. Lee, "Acquisition Performance of a DS/CDMA System in a Mobile Satellite Environment," *IEICE Trans. Commun.,* Vol. E80-B, No. 1, 1997, pp. 40–48.

[45] Restrepo, J., and G. Maral, "Providing Appropriate Service Quality to Fixed and Mobile Users in a Non-GEO Satellite-Fixed Cell System," in *Mobile and Personal Satellite Communications 2,* F. Vatalaro and F. Ananasso, eds, London: Springer, 1996, pp. 79–96.

[46] Ramesh, R., "Availability Calculations for Mobile Satellite Communication Systems," *Proc. 46th IEEE Vehic. Technol. Conf.,* Atlanta, 1996, pp. 1033–1037.

[47] Widrow, B., et al., "Adaptive Antenna Systems," *Proc. IEEE,* Vol. 55, 1967, pp. 21–43.

[48] Compton, R. T., Jr., "An Adaptive Array in a Spread-Spectrum Communication System," *Proc. IEEE,* Vol. 66, 1978, pp. 289–298.

[49] Swales, S. C., et al., "The Performance Enhancement of Multibeam Adaptive Base Station Antennas for Cellular Land Mobile Radio Systems," *IEEE Trans. Vehic. Technol.,* Vol. 39, No. 1, 1990, pp. 56–67.

[50] Anderson, S., et al., "An Adaptive Array for Mobile Communication Systems," *IEEE Trans. Vehic. Technol.,* Vol. 40, No. 1, 1991, pp. 230–236.

[51] Kohno, R., H. Imai, and S. Pasupathy, "Combination of an Adaptive Antenna Array and a Canceller of Interference for Direct-Sequence Spread-Spectrum Multiple-Access System," *IEEE J. Select. Areas Commun.,* Vol. 8, No. 4, 1990, pp. 675–682.

[52] Ogawa, Y., Y. Nagashima, and K. Itoh, "An Adaptive Antenna System for High Speed Digital Mobile Communications," *IEICE Trans. Commun.,* Vol. E75-B, No. 5, 1992, pp. 413–421.

[53] Kuroiwa, N., R. Kohno, and H. Imai, "Design of a Diversity Receiver Using an Adaptive Array Antenna," *IEICE Trans.*, Vol. J73-B-II, No. 11, 1990, pp. 755–763.

[54] Naguib, A. F., A. Paulraj, and T. Kailath, "Capacity Improvement With Base-Station Antenna Arrays in Cellular CDMA," *IEEE Trans. Vehic. Technol.*, Vol. 43, No. 3, 1994, pp. 691–698.

[55] Sumino, H., et al., "The Receiver With an Adaptive Array Antenna and Satellite Diversity for Low Earth Orbital Multiple Satellite Communication Systems," in *Mobile and Personal Satellite Communications 2*, F. Vatalaro and F. Ananasso, eds, London: Springer, 1996, pp. 180–194.

Selected bibliography

Litva, J., T. K-Y. Lo, *Digital Beam Forming in Wireless Communications*, Norwood, MA: Artech House, 1996.

List of Acronyms

ACTS	Advanced Communication Technology Studies (in Europe; formerly RACE)
AGC	automatic gain control
AM	amplitude modulation
AMPS	American Mobile Phone System
AMSC	American Mobile Satellite Corporation
AOR	Atlantic Ocean region
APR	automatic position reporting
ARTEMIS	Advanced Research and Technology Mission
ATDMA	advanced TDMA mobile access
ATPC	adaptive transmit permission control
AWGN	additive white Gaussian noise
BCH	Bose-Chaudhuri-Hocquenghem (coding)
BPSK	binary phase shift keying
BS	broadcasting satellite (also base station)

CCIR	International Radio Consultative Committee (now ITU-R)
CDMA	code division multiple access
COMSAT	Communications Satellite Corporation
CS	control station
DAB	digital audio broadcasting
DAMA	demand assignment multiple access
DBS	direct broadcasting satellite
DCA	dynamic channel allocation
DS/SSA	direct-sequence spread-slotted Aloha
DECT	Digital European Cordless Telecommunications
DL	downlink (also forward link)
DS	direct sequence
DTS	dense-traffic satellite
EIRP	equivalent isotropically radiated power
EMS	European mobile satellite
ESA	European Space Agency
ETSI	European Telecommunication Standards Institute
EUTELSAT	European Telecommunication Satellite organization
FCA	fixed channel allocation
FCC	Federal Communications Commission
FDD	frequency division duplexing
FDM	frequency division multiplexing
FDMA	frequency division multiple access
FES	fixed Earth station
FM	frequency modulation
FPLMTS	Future Public Land Mobile Telecommunication Systems
FSK	frequency shift keying
GEO	geostationary Earth orbit (also GSO: geostationary satellite orbit)
GPS	global positioning system
GSM	global system for mobile communications
HEO	highly elliptic orbit
HIO	highly inclined orbit
ICO	intermediate circular orbit
IF	intermediate frequency

IN	intelligent network
INMARSAT	International Maritime telecommunication Satellite organization
INTELSAT	International Telecommunication Satellite organization
ISDN	Integrated Services Digital Network
ISL	intersatellite link
ITU	International Telecommunication Union
ITU-R	Radio communication sector of ITU (formerly CCIR)
JMPS	Japanese Mobile Phone System
LAN	local area network
LEO	low Earth orbit
LLM	L -band land mobile
MA	multiple access
MAI	multiple access interference
MEO	medium Earth orbit
MIO	multistationary inclined orbits
MONET	Mobile Network
MS	mobile satellite (also mobile station)
MSC	mobile services switching center
MSS	mobile satellite services
NASA	National Aeronautics and Space Administration (U.S.)
NASDA	National Space Development Agency (Japan)
NCS	network control station
ORBCOMM	Orbital Communications Corporation
PABX	private automatic branch exchanger
PAC	packet admission control
PBS	Public Broadcasting Service
PBX	private branch exchange
PCM	pulse code modulation
PCN	personal communication network
PCS	personal communication services
PHS	personal handy phone system (formerly PHP)
PLL	phase locked loop
PN	pseudo-noise
PRMA	packet reservation multiple access
PSK	phase shift keying

PSTN	public switched telephone network
PTT	post, telephone, and telegraph
QPSK	quadrature phase shift keying
RACE	Research and Development in Advanced Communications technologies in Europe (now ACTS)
RDSS	radio determination satellite services
RF	radio frequency
RFI	radio frequency interference
SDMA	space division multiple access
SHF	super high frequency
SIR	signal-to-interference ratio
SNR	signal-to-noise ratio
SS	spread spectrum
SSA	spread-slotted Aloha
SSMA	spread spectrum multiple access
STS	sparse-traffic satellite
SYNCOM	synchronous orbit communications satellite
TDD	time division duplexing
TDM	time division multiplexing
TDMA	time division multiple access
TMI	Telsat Mobile Incorporated
TPC	transmit permission control
UHF	ultra high frequency
UL	uplink (also reverse link)
UMTS	universal mobile telecommunication systems
UPT	universal personal telecommunications
VHF	very high frequency
VITA	Volunteers in Technical Assistance
VSAT	very small aperture terminal
WAN	wide area network
WARC	World Administrative Radio Conference
WATS	wide area telecommunication service

About the Author

ABBAS JAMALIPOUR received his Ph.D. in electrical engineering from Nagoya University, Japan, where he is now an assistant professor in the Department of Information Electronics. During his education, he engaged in different fields of computer and communications engineering; his current interests are in digital wireless communications, satellite communications, mobile communications, computer communication networks, spread-spectrum communications, and traffic and congestion control. He is a member of IEEE and a member of IEEE Communications Society technical committees on Satellite and Space Communications, Personal Communications, and Communications Switching. He is also a member of the International Union of Radio Science, the IEICE, and SITA of Japan.

Mr. Jamalipour received the URSI Young Scientists Award at the 25th General Assembly of the URSI, held in Lille, France, in 1996. He also received the 13th Inoue Research Award for Young Scientists from the Inoue Foundation for Science of Japan in 1996 for his outstanding researches on LEO satellite systems. In 1997, he received the annual Telecommunications Technology Paper Award for his paper published in the February 1995 issue of the *IEEE Journal on Selected Areas in Communications* entitled "Performance of an Integrated Voice/Data System in Nonuniform Traffic Low Earth-Orbit Satellite Communication Systems" from the Telecommunications Advancement Foundation of Japan. His name has been selected for inclusion in the 15th edition of the *Marquis Who's Who in the World* for his outstanding contributions to international scientific activities.

Index

Abramson, Norman, 119, 124, 129
Adaptive array antennas, 248–50
   defined, 248–49
   employing, 249–50
   illustrated, 250
   structure of, 249
   uses, 249
   See also Antennas
Adaptive TPC (ATPC), 186–87, 214–19
   defined, 215
   method 1, 215–16
   method 2, 217–18
   performance of, 217–18
   See also Transmit permission control
      (TPC)
Additive white Gaussian noise
      (AWGN), 126
Aloha, 100
   capacity, 123

CDMA, 63, 124, 125–27
channel packet rate, 120–21
channel throughput, 120–21
collisions, 123
creation of, 119
defined, 59, 119
in FDMA, 124
multiple-access scheme, 119–23
offered traffic loads and, 180
packet collisions, 123
packet flow, 120
packets, spreading, 123–29
pure, 121
slotted, 63, 121
spread, 124, 128–29
spread-slotted, xv, 34, 63, 117–53
system comparison, 122
in TDMA, 124
unslotted, 63–64, 121

263

Altitudes
  Kepler's third law and, 11
  minimum number of orbits and, 40
  satellite system comparison by, 16
American Mobile Satellite Corporation
      (AMSC), 21
Analog systems, 79–96
  SIR, 82–91
  traffic assignment control, 91–96
  traffic modeling, 79
  See also CDMA
Antennas
  adaptive array, 248–50
  boresight, 242, 243
  gain, 242–45
  IRIDIUM, 23
  isotropic, 243
  ODYSSEY, 26
  spot-beam, 47, 67
Aperture efficiency, 243
Average delay, 186
  calculating, 219
  comparing, 204–5
  defined, 130, 201
  effect on TPC scheme, 206
  expressing, 205
  graphs of, 131
  increase in, 219
  normalized, 203, 206
  performance, 207
  performance analysis, 187
  performance of TPC scheme, 203
  throughput and, 131, 203, 208
  throughput performance
      comparison, 206
  TPC performance, 201–8
  use of, 130
Average normalized total
      throughput, 201–2
  defined, 201
  TPC effect on, 201, 202

Bent-pipe network architecture, 25, 26

defined, 25
illustrated, 26
Big-LEO, 22–27
  ARIES, 26–27
  defined, 19
  GLOBALSTAR, 25
  IRIDIUM, 22–25
  ODYSSEY, 25–26
  proposed systems, 22–27
  services, 22
  See also LEO satellite systems
Binary phase shift keying (BPSK)
  direct-sequence (DS/BPSK), 138
  modulation, 53
Bit error probability, 137
Boresight, 242, 243
Bose-Chaudhuri-Hocquenghem (BCH), 97

Capacity, system, 166
Capture probability, 191, 195
  improving, 192
  interference and, 192–93
  low, 198
CDMA, *xv*, 48
  advantages, *xiv*, 62–63
  application of, 77–113
  carrier frequency use, 77
  circuit-mode traffic, 63, 97
  code multiplexing efficiency, 103
  defined, 59
  direct sequence
      (DS/CDMA), 60, 79, 82
  direct-sequence signals, 125–28
  frequency reuse, 50
  GLOBALSTAR and, 25
  LEO satellite systems, 64
  on uplinks, 78
  packet-mode traffic, 63, 97
  peak channel speed, 97
  performance evaluation, 79–82
  schemes, 59, 62, 96
  signals, 59
  SIR and, 82–91

spatial reuse in, 103
  See also Multiple-access schemes
CDMA Aloha, 63, 124
  packets in, 127
  slotted, 127
  See also Aloha
Cells
  cluster of, 50
  defined, 47
  inserted within footprints, 48
Chip interval, 61
Closed-loop power control, 169, 224, 238
  defined, 169
  effectiveness, 224
  See also Power control
Code division multiple access. *See* CDMA
Collisions, 123, 202
  avoiding, of same packets, 202
  weakening effect of, 123
Communications
  with LEO satellites, 33–73
  methods, 3–5
Communications satellites, 2–7
  activities, 5
  coverage, 6
  defined, 4
  features, 6
  history of, 5–7
  preliminary issues, 2–5
Contention systems, 120
Conventional power control, 168–69
  closed-loop, 169
  open-loop, 168–69
  radiated power, 173
  user connection, 171
  See also Modified power control; Power
    control
Coverage area, 70
  defined, 68
  double, 70
  service area/interference area,
    relationship with, 72

Data packet transmission, 98
Delay performance
  average, 201
  improving, 203
  See also Average delay
Dense traffic satellite (DTS), 89, 104
  nonuniform traffic and, 147
  performance, 147–49
  performance, improving, 94
  service area, 94
  service area users, 147
  signal quality improvement of, 174
  SIR characteristic changes, 92
  SIR characteristics, 90
  SIR characteristics as function of traffic
    nonuniformity, 106, 107, 108
  transmitting power, increasing, 94
  See also Sparse traffic satellite (STS)
Density function, 189–90
Directivity, 242
Direct sequence CDMA
    (DS/CDMA), 60, 79, 82
Direct-sequence-spread-slotted Aloha
    (DS/SSA), 117, 190
  capacity and, 190
  conventional, 190
  modifying, 190–91
  See also Spread-slotted Aloha
Direct-to-multipath signal power ratio, 210
Distribution of users, 132–34
Doppler shift effect, 51–55
  canceling, 54–55
  cause of, 51
  defined, 51
  frequency shift from, 53
  illustrated, 52
  in LEO satellite systems, 34
  normalized, 53
  numerical examples, 54
  schematic diagram, 53
Double coverage area, 70
Dynamic nonuniform traffic, 108–13

Dynamic nonuniform traffic (continued)
   modified power control
      scheme, 110–13
   simulation model, 109–10
   See also Traffic nonuniformity

Effective aperture area, 243
Effective isotropically radiated power
      (EIRP), 243
ELLIPSO, 27
Error-correcting code, 145
Error function, 144, 145
European mobile satellite (EMS), 21

Fading
   absence of, 210, 211, 240
   effect of, 213, 241
   level of, 211
   model, 186, 230
   multipath, 209
   on uplink, 191, 238
   Rayleigh, 84, 140, 210, 238
   variation compared to bit rate, 212
Fading channels, 209–14
   model, 209–12
   throughput analysis, 186
   throughput performance, 213
   throughput performance
      comparison, 231
   uplink, 212
   See also Transmit permission
      control (TPC)
FDMA
   Aloha in, 124
   defined, 57
   frame structure, 58
   guard band, 59
   use of, 57
   VSATs and, 62
   See also Multiple-access schemes
Federal Communications Commission
      (FCC), 21
Fiber-optic cable, 18–19

Footprints, 37, 38, 46
   cells inserted within, 48
   GLOBALSTAR satellite, 51
   IRIDIUM satellite, 51
   partitioning, 47, 48
Frequency division duplexing (FDD), 27
Frequency division multiple access.
      See DMA
Frequency division multiplexing (FDM), 48
Frequency reuse, 247
   CDMA, 50
   seven-frequency pattern, 247
   TDMA, 50
   three-frequency pattern, 247
Future public land mobile
      telecommunication system
      (FPLMTS), 1

Gain
   defined, 242
   general expression, 242–43
   processing, 60
   spot-beam antenna, 243–45
   See also Antennas
Gaussian noise, 100–101
Geographic traffic nonuniformity, 34, 65
GEO satellites, 6
   comparison, 14–15, 17
   coverage requirement, 16
   example, 10–11
   frequency spectrum, 11
   lack of coverage, 13–14
   launch cost of, 14
   mobile communications with, 19
   problems with, 12–14
   propagation delay, 12
   propagation loss, 12–13
Geostationary Earth orbit satellites.
      See GEOs
Global communications network, 188
GLOBALSTAR, 25
   downlink/uplink frame structures, 49
   satellite footprint, 51

satellite requirements, 38

Hand-offs, 41–43
  average number of, 42
  defined, 42
  LEO satellite systems, 42–43
  mechanism, 43
Highly elliptical orbits (HEO), 16

Improved Gaussian approximation, 137
INMARSAT
  defined, 20
  INMARSAT A, 20
  INMARSAT B, 20
  INMARSAT C, 20
  mini-M, 21
  standards, 20
  terminals, 20
Integrated voice/data systems, 96–113
  dynamic nonuniform traffic
      concepts, 108–13
  packet formatting, 98
  performance measurement, 103–8
  simulation environment, 101–3
  system considerations, 96–99
  traffic model extension, 99–101
Intercell interference, 160–64, 180
Interference
  cancellation methods, 249
  capture probability and, 192–93
  decreasing level of, 158
  effects, 223, 228
  intercell, 163–64, 180
  intracell, 163–64
  multiple-access, 163, 185, 195,
      230, 250
  power, 181, 198
  reducing, 232
  shadowing and, 231
  simultaneously transmitted packets, 191
  from users outside service area, 161
Interference area, 70, 136
  defined, 69

illustrated, 69
radius, 226
service area/coverage area, relationship
    with, 72
Intermediate circular orbit (ICO), 27, 210
International Maritime Telecommunication
    Satellite Organization. *See*
    INMARSAT
International Telecommunications Satellite
    Organization (INTELSAT), 7
Intersatellite links (ISL), 43–46
  defined, 43
  Earth gateway stations and, 44
  GEO satellites and, 43
  inter-orbit, 45
  intra-orbit, 45
  IRIDIUM, 46
  LEO, 44–46
  types of, 45
  use proposals, 44
Intracell interference, 160–64
IRIDIUM, 22–25, 34
  antennas, 23
  coverage requirement, 16, 23, 38
  defined, 22–23
  features, 23
  ISLs, 46
  orbit constellation, 24
  processing facilities, 24
  satellite footprint, 51
  satellite mass, 23
  TDMA frame structure, 49
  transceivers, 23
Isotropic area, 243

Kepler's first law, 8
Kepler's second law, 8, 11
Kepler's third law, 8–10, 11

LEO satellite systems, *xiii*, 14
  big-LEO, 19, 22–27
  CDMA, 64
  communications with, 33–73

LEO satellite systems (continued)
  comparison, 17
  cost of, 17
  Doppler shift effects, 34
  dynamic features of, 108–9
  effective, 248
  future personal mobile communication
     systems, 18
  geometrical considerations, 36
  hand-off, 41–43
  implementation of, 34
  intersatellite links, 43–46
  issues in, 55–67
  little-LEO, 19, 21–22
  mobile communication with, 19
  modeling, 34–35, 67–72
  multibeam, 241–48
  multicell, 51
  multiple-access scheme selection, 56–64
  network topology, 132
  orbits, 35–41
  performance, 132
  performance, analyzing, 68
  preliminary issues in, 35–55
  propagation delay, 17, 187
  propagation loss, 17
  satellite requirements, 35–41
  service areas, xiv, 65
  services, 64–65
  spot beams, 46–51
  traffic considerations, 64–67
Little-LEO, 21–22
  defined, 19
  frequency spectrum, 22
  licenses, 21
  See also LEO satellite systems
Local area networks (LANs), 187
Log-normal distributed random
     variable, 239
Low Earth orbital satellites. See LEO
    satellite systems

Markov M-state channel model, 210

Medium Earth orbit satellites
    (MEO), 14, 17
Microcellular systems, 41
Mixture density, 21
Mobile satellite systems, 12–27
  big-LEO, 19, 22–27
  GEO, 19
  LEO, 19
  little-LEO, 19, 21–22
  in operation, 20–21
  orbital dynamics of, 7–11
  orbit selection, 12–18
  satellite requirements in, 40
Modeling, 67–72
  during satellite movement, 78
  three-dimensional, 70
  traffic, 79–82
  two-dimensional, 70–71
Modified power control, 110–13
  defined, 158
  goal of, 158
  numerical examples, 172–79
  performance evaluation, 172, 175
  radiated power, 173
  realization of, 179–80
  service areas and, 172
  in spread-slotted Aloha, 157–81
  throughput characteristics and, 176
  use of, 173
  user connection, 171
  See also Conventional power control;
Power control
Multibeam LEO satellites, 241–48
Multipath fading
  defined, 209
  modeling, 239
  power variations, 240
  See also Fading
Multiple-access capability, 140
  medium value of, 197
  small value of, 197
Multiple-access schemes
  classification of, 56–61

comparative study, 61
defined, 56–57
in satellite systems, 61–64
selection of, 56–64
spread-spectrum, 60–61
See also CDMA; FDMA; TDMA

Natural service area, 71
Near-far problem, 83, 236–37
defined, 168
in direct-sequence systems, 168
severe, 168
solving, 236–37
See also Power control
Nonfading channel, 187–208
throughput performance
comparison, 231
See also Transmit permission control
(TPC)
Nonuniformity distribution, 104
Nonuniform traffic models, 79–82
examples, 80
shapes, 80
SIR and, 88–91
typical shape of, 81
See also Traffic nonuniformity
Normalized average delay, 203
defined, 203
normalized total throughput vs., 206–7
Normalized far-field radiation pattern, 244
Normalized throughput, 134, 146, 148
average, 201
calculation of, 198
as function of composite packet
transmission, 233–35
of individual satellites, 177
maximized, 148, 149, 175
maximized values, 176
normalized average delay vs., 206–7
results, 176
total, 150, 151
characteristics, 179
comparison, 166, 167, 178

effect on TPC, 201
See also Throughput
Normal-shape function, 226
NSTAR, 21

ODYSSEY, 25–26
antennas, 26
defined, 25–26
orbit inclination, 26
satellite requirements, 38
Open-loop power control, 168–69, 238
defined, 168
perfect, 169
See also Power control
Optimum control, 91–93
OPTUS, 21
Orbits
altitudes, 40
circular, 15
comparison of, 14–18
elliptical, 15
LEO, 35–41
minimum number of, 40, 41
polar, 40
selection of, 12–18
Organization, this book, *xv–xvii*
Oribital Communications Corporation
(ORBCOMM), 21
Overlapping area, 150

Packet admission control (PAC), 224–35
concepts, 232–33
constraints, 232–33
defined, 224–25
heavy-traffic performance, 228–31
packet success probability, 235
performance, 233–35
system and traffic models, 225–28
TPC performance vs., 235
Packets
acknowledgment, 205
Aloha, spreading, 123–29
capture probability, 191

Packets (continued)
  CDMA Aloha, 124
  CDMA generation example, 127
  collisions, 123, 202
  flow in Aloha channel, 120
  lost, 205
  retransmission of, 131, 202
  simultaneous throughput, 136
  spread Aloha, 124
  spread Aloha generation example, 128
  tagged, 191, 195, 196
  transmission in PAC method, 225
Packet success probability, 101, 136–44
  calculating, 136–44
  conditional, 138, 145
  improving, 203
  low, 199
  nonperfect capture, 137
  nonzero, 124
  numerical examples, 144–51
  PAC, 235
  perfect capture, 137
  in service areas, 139
  slotted/unslotted, 137
  in TPC scheme, 187
  unconditional, 138–39
  use of, 130
Perigee, 8
Personal communication networks
(PCNs), xiii, xiv
Personal communications services (PCS), 1
Personal handy phone system (PHS), 2
Phase locked loop (PLL), 55
Picocellular systems, 41, 42
Power control, 83, 224, 236–41
  closed-loop, 224, 238
  conventional, 168–69
  imperfections in, 238–41
  implementation, 224, 237–38
  importance of, 169
  modified scheme, 110–13, 157–81
  near-far problem and, 236–37
  on reverse links, 237

open-loop, 238
  selection parameters, 169
Predefined signals, 189
Probability density function, 233
Probability distribution function, 233
Processing gain, 60
Propagation delay
  in average delay, 130
  elevation angle relationship with, 13
  GEO, 12–13
  LEO, 17, 187
  performance, improving, 203
  total, 203
Propagation loss, 192
  GEO, 12–13
  LEO, 17
  maximum, 196
Pure Aloha, 121

Quadrature phase shift keying (QPSK), 48

Radiated powers comparison, 173
Radiation patterns
  far-field, 244
  modeling, 244
  normalized far-field, 246
  spot-beam, 245
Radio determination satellite services
      (RDSS), 22
Rayleigh fading, 84, 140
  selective, 210
  variations due to, 238
  See also Fading
Reflection method, 3
Retransmissions
  delay, 202, 205
  increasing number of, 202–3
  packet, 131, 202

Satellites
  altitude of, 11
  angular velocity of, 9
  footprints, 37, 38, 46

rotating path of, 8
Service areas, 70
  circular, hexagon, 193
  configuration of, 227
  coverage area/interference area
    relationship with, 72
  defined, 68
  of DTS, 94
  equal-size, 171
  as hexagons, 227
  modified power control and, 172
  natural, 71
  nonoverlapping, 193
  packet success probabilities in, 139
  reduction in, 195
  size of, changing, 171
  TPC method application and, 193–94
  traffic ratio in, 83
  uniform configuration of, 170
  upper bound for, 69
Shadowing, 84, 140
  absence of, 210, 240
  defined, 209
  effect of, 230
  interference power and, 231
  modeling, 239
  time-share of, 210
Signal-to-interference ratio
    (SIR), 78, 82–91
  calculation of, 79
  CDMA and, 82–87
  changes at main satellite, 111, 112
  characteristics, changes at DTS/STS, 92
  characteristics, changes in, 88
  at DTS/STS, 90
  instantaneous levels of, 164
  measurement, 85
  required, 85
  satellite position and, 87–88
  in spread-spectrum systems, 101
  traffic nonuniformity and, 88–91
Simulation
  environment, 101–3

model, 101–2, 109–10
  parameters, 102–3
  start point of, 110
Slotted Aloha, 63, 121
  packet success probability
    calculation, 137
  throughput, 122, 130
  See also Aloha
Sparse traffic satellite (STS), 89, 104
  nonuniform traffic and, 147
  performance, 147–49
  service area users, 147
  SIR characteristic changes, 92
  SIR characteristics, 90
  See also Dense traffic satellite (DTS)
Spot-beam antennas, 47, 67
  configuration, 244
  gain, 243–45
  performance of, 245–48
  See also Antennas
Spread Aloha, 124, 128–29
  offered traffic loads and, 180
  packet generation example, 128
  packets, 128
  slotted/unslotted, 129
  See also Aloha
Spread-slotted Aloha systems,
    xv, 34, 117–53
  defined, 63
  direct-sequence (DS/SSA), 117, 190
  employing, 118, 130–44
  modified power control in, 157–81
  transmit permission control
    for, 185–220
  See also Aloha
Spread spectrum, 63, 73
  multiple access (SSMA), 60–61
  network-analysis level, 100
  signal-to-interference ratio and, 101
  system example, 125
STARNET, 21
Step function, 145
Stochastic model, 223

System capacity, 166

Tagged packets, 191, 195, 196
TDMA, 24
  advantages of, 61–62
  Aloha in, 124
  defined, 57
  frames, 25, 48
  frame structure, 49, 58
  guard time, 59
  seven-frequency reuse pattern, 50
  use of, 57
  See also Multiple-access schemes
Throughput
  analysis, 131, 134–36
  average delay and, 131, 203, 208
  average value, 135
  comparisons, 229
  defined, 120
  in fading channel, 213
  fading/nonfading channel
    comparison, 231
  graphs, 131
  instantaneous, averaging, 201
  low, 230
  maximized normalized, 148, 149, 175
  maximum value, 166
  nonworst case performance, 164–68
  normalized, 134, 146, 148
  normalized total, 150, 151, 166, 167,
    178, 179
  PAC, 233–35
  performance comparison, 159–62
  performance worst case, 159–64
  simultaneous packet, 136
  single-satellite, 195
  slotted Aloha, 122, 130
  total, 195, 196
  TPC performance, 195–202
  use of, 130
Time division multiple access. *See* TDMA
Time division multiplexing (TDM), 24
Traffic, 64–67

circuit-mode, 63, 97
densities assignment, 66
geographic nonuniformity, 34, 65
integrated voice/data, 96–113
model extension, 99–101
modeling, 79–82
packet-mode, 63, 97
ratio in service area, 83
uniform, 91
uniform distribution, 82, 163
See also Nonuniform traffic model;
  Traffic nonuniformity
Traffic assignment control, 91–96
  optimum, 91–93
  optimum capability
    measurement, 93–96
Traffic loads
  heavy offered, 177, 180, 199
  high, 167
  high offered, 201
  light offered, 213
  low offered, 177
  offered, 180
  offered to CDMA channel, 100
  peak of, 170, 171
Traffic nonuniformity, 181
  distribution, 109, 230
  DTS/STS and, 147
  dynamic, 108–13
  large, 174
  level change, 104
  measure, 110
  models, 79–82
  parameter relationship, 105
  SIR characteristics at DTS/STS as
    function of, 106, 107, 108
  system performance and, 147, 150
  TPC scheme and, 208
  traffic loads and, 167
Transmit permission control
  (TPC), 185–220
  adaptive, 186–87, 214–19
  application of, 213

average normalized total throughput
    effect on, 201
average waiting time, 204
defined, 186
effect on average delay, 206
fading channel, 209–14
improvements achieved by, 198
instruction, 192
low elevation angles and, 209
nonfading channel, 187–208
numerical examples, 212–14
packet success probability, 187
PAC performance vs., 235
physical representation of, 193
realization of, 192
throughput performance, 195–202
throughput performance effect over
    dense traffic area, 215
uses, 186

Uniform traffic, 91
    distribution, 82, 163, 230
    service area size and, 91
Universal mobile telecommunication
    services (UMTS), 1
Universal personal telecommunications
    (UPT), 1
Unslotted Aloha, 63–64, 121, 137

Van Allen radiation belts, 14, 15
Very small aperture terminals (VSATs), 62
Volunteers in Technical Assistance
    (VITASAT), 21

Worst case performance, 159–64

The Artech House Mobile Communications Series

John Walker, Series Editor

Advanced Technology for Road Transport: IVHS and ATT,
Ian Catling, editor

An Introduction to GSM, Siegmund M. Redl, Matthias K. Weber,
Malcolm W. Oliphant

CDMA for Wireless Personal Communications, Ramjee Prasad

Cellular Communications: Worldwide Market Development,
Garry A. Garrard

Cellular Digital Packet Data, Muthuthamby Sreetharan, Rajiv Kumar

Cellular Mobile Systems Engineering, Saleh Faruque

Cellular Radio: Analog and Digital Systems, Asha Mehrotra

Cellular Radio: Performance Engineering, Asha Mehrotra

Cellular Radio Systems, D. M. Balston, R. C. V. Macario, editors

Digital Beamforming in Wireless Communications, John Litva and
Titus Kwok-Yeung Lo

GSM System Engineering, Asha Mehrotra

Handbook of Land-Mobile Radio System Coverage, Garry C. Hess

Introduction to Wireless Local Loop, William Webb

*Introduction to Radio Propagation for Fixed and Mobile
Communications,* John Doble

Land-Mobile Radio System Engineering, Garry C. Hess

Low Earth Orbital Satellites for Personal Communication Networks,
Abbas Jamalipour

Mobile Antenna Systems Handbook, K. Fujimoto, J. R. James

*Mobile Communications in the U.S. and Europe: Regulation,
Technology, and Markets,* Michael Paetsch

Mobile Data Communications Systems, Peter Wong, David Britland

Mobile Information Systems, John Walker, editor

Personal Communications Networks, Alan David Hadden

RF and Microwave Circuit Design for Wireless Communications, Lawrence E. Larson, editor

Smart Highways, Smart Cars, Richard Whelan

Spread Spectrum CDMA Systems for Wireless Communications, Savo G. Glisic, Branka Vucetic

Transport in Europe, Christian Gerondeau

Understanding GPS: Principles and Applications, Elliott D. Kaplan, editor

Universal Wireless Personal Communications, Ramjee Prasad

Vehicle Location and Navigation Systems, Yilin Zhao

Wireless Communications for Intelligent Transportation Systems, Scott D. Elliott and Daniel J. Dailey

Wireless Communications in Developing Countries: Cellular and Satellite Systems, Rachael E. Schwartz

Wireless Data Networking, Nathan J. Muller

Wireless: The Revolution in Personal Telecommunications, Ira Brodsky

For further information on these and other Artech House titles, including previously considered out-of-print books now available through our In-Print-Forever™ (IPF™) program, contact:

Artech House   Artech House
685 Canton Street   Portland House, Stag Place
Norwood, MA 02062   London SW1E 5XA England
781-769-9750   +44 (0) 171-973-8077
Fax: 781-769-6334   Fax: +44 (0) 171-630-0166
Telex: 951-659   Telex: 951-659
e-mail: artech@artech-house.com   e-mail: artech-uk@artech-house.com

Find us on the World Wide Web at: www.artech-house.com